Memoirs of the American Mathematical Society
Number 373

Minking Eie

Dimension formulae for the vector spaces of Siegel cusp forms of degree three (II)

Published by the
AMERICAN MATHEMATICAL SOCIETY
Providence, Rhode Island, USA

November 1987 · Volume 70 · Number 373 (first of 6 numbers)

MEMOIRS of the American Mathematical Society

SUBMISSION. This journal is designed particularly for long research papers (and groups of cognate papers) in pure and applied mathematics. The papers, in general, are longer than those in the TRANSACTIONS of the American Mathematical Society, with which it shares an editorial committee. Mathematical papers intended for publication in the Memoirs should be addressed to one of the editors:

Ordinary differential equations, partial differential equations, and applied mathematics to JOEL A. SMOLLER, Department of Mathematics, University of Michigan, Ann Arbor, MI 48109

Complex and harmonic analysis to ROBERT J. ZIMMER, Department of Mathematics, University of Chicago, Chicago, IL 60637

Abstract analysis to VAUGHAN F. R. JONES, Department of Mathematics, University of California, Berkeley, CA 94720

Classical analysis to PETER W. JONES, Department of Mathematics, Box 2155 Yale Station, Yale University, New Haven, CT 06520

Algebra, algebraic geometry, and number theory to DAVID J. SALTMAN, Department of Mathematics, University of Texas at Austin, Austin, TX 78713

Geometric topology and general topology to JAMES W. CANNON, Department of Mathematics, Brigham Young University, Provo, UT 84602

Algebraic topology and differential topology to RALPH COHEN, Department of Mathematics, Stanford University, Stanford, CA 94305

Global analysis and differential geometry to JERRY L. KAZDAN, Department of Mathematics, University of Pennsylvania, E1, Philadelphia, PA 19104-6395

Probability and statistics to RONALD K. GETOOR, Department of Mathematics, University of California at San Diego, La Jolla, CA 92093

Combinatorics and number theory to RONALD L. GRAHAM, Mathematical Sciences Research Center, AT&T Bell Laboratories, 600 Mountain Avenue, Murray Hill, NJ 07974

Logic, set theory, and general topology to KENNETH KUNEN, Department of Mathematics, University of Wisconsin, Madison, WI 53706

All other communications to the editors should be addressed to the Managing Editor, LANCE W. SMALL, Department of Mathematics, University of California at San Diego, La Jolla, CA 92093.

PREPARATION OF COPY. Memoirs are printed by photo-offset from camera-ready copy prepared by the authors. Prospective authors are encouraged to request a booklet giving detailed instructions regarding reproduction copy. Write to Editorial Office, American Mathematical Society, Box 6248, Providence, RI 02940. For general instructions, see last page of Memoir.

SUBSCRIPTION INFORMATION. The 1987 subscription begins with Number 358 and consists of six mailings, each containing one or more numbers. Subscription prices for 1987 are $227 list, $182 institutional member. A late charge of 10% of the subscription price will be imposed on orders received from nonmembers after January 1 of the subscription year. Subscribers outside the United States and India must pay a postage surcharge of $25; subscribers in India must pay a postage surcharge of $43. Each number may be ordered separately; *please specify number* when ordering an individual number. For prices and titles of recently released numbers, see the New Publications sections of the NOTICES of the American Mathematical Society.

BACK NUMBER INFORMATION. For back issues see the AMS Catalogue of Publications.

Subscriptions and orders for publications of the American Mathematical Society should be addressed to American Mathematical Society, Box 1571, Annex Station, Providence, RI 02901-9930. *All orders must be accompanied by payment.* Other correspondence should be addressed to Box 6248, Providence, RI 02940.

MEMOIRS of the American Mathematical Society (ISSN 0065-9266) is published bimonthly (each volume consisting usually of more than one number) by the American Mathematical Society at 201 Charles Street, Providence, Rhode Island 02904. Second Class postage paid at Providence, Rhode Island 02940. Postmaster: Send address changes to Memoirs of the American Mathematical Society, American Mathematical Society, Box 6248, Providence, RI 02940.

The paper used in this journal is acid-free and falls within the guidelines established to ensure permanence and durability. ♾

TABLE OF CONTENTS

ABSTRACT

The well known Selberg trace formula reduces the problem of calculating the dimension of cusp forms of Siegel upper-half plane, when the fundamental domain is not compact but has finite volume, to the evaluation of certain integrals combining with special values of certain zeta functions. In this paper, we shall obtain explicit dimension formulae for cusp forms of degree three with respect to the full modular group $Sp(3, \mathbf{Z})$ and its principal congruence subgroups by a long computation.

AMS Subject Classification: Primary 10D20

This work was supported by Academia Sinica and N.S.F. (NSC74-0208-M001-03) of Taiwan, R.O.C.

Library of Congress Cataloging-in-Publication Data

Eie, Minking, 1952–

Dimension formulae for the vector spaces of Siegel cusp forms of degree three (II).

(Memoirs of the American Mathematical Society, ISSN 0065-9266; no. 373)

"November 1987."

"Volume 70, number 373 (first of 6 numbers)."

Bibliography: p.

1. Cusp forms (mathematics) 2. Selberg trace formula. 3. Integrals. I. Title. II. Title: Siegel cusp forms of degree three (II). III. Series.

QA3.A57 no. 373 510 s [512′.72] 87-25477

[QA243]

ISBN 0-8218-2436-8

NOTATION

1. $\mathbf{Z}$, $\mathbf{Q}$, $\mathbf{R}$, $\mathbf{C}$: ring of rational integers and the fields of rational numbers, real numbers and complex numbers respectively.
2. $M_n(Z)$, $M_n(R)$, $M_n(\mathbf{C})$: rings of $n \times n$ matrices over Z, $\mathbf{R}$, $\mathbf{C}$ respectively.
3. $GL_n(Z)$, $GL_n(R)$: general linear groups over $\mathbf{Z}$, $\mathbf{R}$ respectively.
4. $SL_n(Z)$, $SL_n(R)$: special linear groups over $\mathbf{Z}$, $\mathbf{R}$ respectively.
5. $U(n)$: group of $n \times n$ unitary matrices;

$$U(n) = \{U \in M_n(\mathbf{C}) \mid U^{-1} = {}^t\bar{U}\} .$$

6. $Sp(n, \mathbf{R})$: the real symplectic matrices of degree n; specifically,

$$Sp(n, \mathbf{R}) = \left\{M \in M_{2n}(\mathbf{R}) \mid {}^tMJM = J,\ J = \begin{bmatrix} 0 & E_n \\ -E_n & 0 \end{bmatrix}\right\}.$$

Here E_n is the identity of matrices ring $M_n(\mathbf{C})$.

7. $Sp(n, \mathbf{Z}) = Sp(n, \mathbf{R}) \cap M_{2n}(\mathbf{Z})$: the discrete modular subgroup of degree n.
8. $\Gamma_n(N)$: the principal congruence sungroup of level N of $Sp(n, \mathbf{Z})$, specifically,

$$\Gamma_n(N) = \{M \in Sp(n, \mathbf{Z}) \mid M \equiv E_{2n} (\mathrm{mod}\ N)\} .$$

9. H_n: Siegel upper-half space of degree n ; specifically,

$$H_n = \{Z \in M_n(\mathbf{C}) \mid {}^tZ = Z,\ \operatorname{Im} Z > 0\} .$$

10. D_n: the generalized disc of degree n ; specifically,

$$D_n = \{W \in M_n(\mathbf{C}) \mid {}^tW = W,\ E - W{}^t\bar{W} > 0\} .$$

11. $[S, U]$: element of $Sp(n, \mathbf{R})$ of the form $\begin{bmatrix} E & S \\ 0 & E \end{bmatrix}\begin{bmatrix} U & 0 \\ 0 & {}^tU^{-1} \end{bmatrix}$.

12. diag $[a_1, a_2, \ldots, a_n]$ or $[a_1, a_2, \ldots, a_n]$: the diagonal matrix

$$\begin{bmatrix} a_1 & & & & \\ & a_2 & & 0 & \\ & & \cdot & & \\ & 0 & & \cdot & \\ & & & & a_n \end{bmatrix}.$$

13. $\Gamma(s)$: gamma-function; it is defined by

$$\Gamma(s) = \int_0^\infty e^{-t} t^{s-1} dt \quad \text{for} \quad \operatorname{Re} s > 0 .$$

14. $\alpha(k) = [a_0, a_2, \ldots, a_{m-1}] \pmod{2m}$: $\alpha(k) = a_j$ if $k \equiv 2j$ for $j = 0, 1, \ldots, m-1$.

INTRODUCTION

Let H_n be the generalized upper-half plane of degree n and Γ be a subgroup of the symplectic group $Sp(n, \mathbf{R})$, which acts on H_n properly discontinuous (i.e., given two compact subset A and B of H_n, the set $\Gamma_{A,B} = \{\gamma \in \Gamma \mid \gamma(A) \cap B \neq \phi\}$ is finite) on H_n. Denote by $S(k; \Gamma)$ be the vector space of Seigel cusp forms of weight k and degree n with respect to Γ. In other words, $S(k; \Gamma)$ consists of holomorphic function f on H_n satisfying the following conditions:

(1) $f(\gamma(Z)) = \det(CZ+D)^k f(Z)$ for all $\gamma = \begin{bmatrix} A & B \\ C & D \end{bmatrix} \in \Gamma$.

(2) Suppose that $\Sigma\, a(T)[\exp 2\pi i\ (TZ)]$ is the Fourier expansion of f; then $a(T) = 0$ if rank $T < n$. Here the summation is over all half integral matrices T such that $T \geqslant 0$ and $\sigma(TZ) =$ trace of TZ.

The second condition can be replaced by the growth condition as follows:

(2') $(\det \operatorname{Im} Z)^{k/2} |f(Z)|$ is bounded on H_n.

It is well known that $S(k; \Gamma)$ is a finite dimensional vector space. Furthermore, the dimension of $S(k; \Gamma)$ over $\mathbf{C}$ is given by Selberg trace formula as follows [12]:

Received by the editor April 11, 1986.

$$\dim_{\mathbf{C}} S(k;\ \Gamma) = C(k,\ n)\int_F \sum_M K_M(Z,\ \bar{Z})^k dZ$$

when $k \geqslant 2n+3$.

1. $C(k,\ n) = 2^{-n}(2\pi)^{-n(n+1)/2} \prod_{i=0}^{n-1} \Gamma(k - \frac{n-i-1}{2}) \prod_{i=0}^{n-1}[\Gamma(k-n+\frac{i}{2})]^{-1}$,

2. F is a fudamental domain on H_n for Γ ,

3. In the summation M ranges over all matrices $\begin{bmatrix} A & B \\ C & D \end{bmatrix}$ in $\Gamma/\{\pm 1\}$,

4. $K_M(Z,\ \bar{Z}) = (\det \operatorname{Im} Z)\det(\frac{Z-M(\bar{Z})}{2i})^{-1} \det(C\bar{Z}+D)^{-1}$ for $M = \begin{bmatrix} A & B \\ C & D \end{bmatrix} \in \Gamma$,

5. dZ is the symplectic volume defined by

$$dZ = (\det Y)^{-(n+1)} dXdY \quad \text{if} \quad Z = X+iY \ .$$

Our main interest in this paper is to compute explicitly $\dim_{\mathbf{C}} S(k;\ \Gamma)$ via Selberg trace formula when $\Gamma = Sp(3,\ \mathbf{Z})$.

As claimed in my previous paper [11], a dimension formula for the vector space of Siegel cusp forms of degree three with respect to $Sp(3,\ \mathbf{Z})$ can be obtained once the conjugacy classes of $Sp(3,\ \mathbf{Z})$ are given explicitly. However, the number of conjugacy classes in $Sp(3,\ \mathbf{Z})$ are so large that one cannot expect to get a correct formula without making any mistake in the computation of more than 300 contributions. Fortunately, we observe that $\dim_{\mathbf{C}} S(k; Sp(3, \mathbf{Z}))$ is a finite sum of $P(k)C(k)$ with $P(k)$ being an integral divisor of $(2k-2)(2k-3)(2k-4)^2 \times (2k-5)(2k-6)$ such as $(2k-2)(2k-4)^2(2k-6), (2k-3)(2k-5)(2k-4)$ and $C(k)$ being a constant or a period function in k such as $(-1)^k$, $\cos(2k-2)\pi/3$, $\sin(k-2)\pi/3$. After selected [14] concributions (we call these contributions the main terms) from the dimension formula, we found that the sum of the remaining terms appears to be the form

$$C_1(k)(2k-4)^2 + C_2(k)(2k-4) + C_3(k)$$

with $C_j(k) = C_j(k+12)$, $j = 1, 2, 3$.

Note that the sum of the main terms and $C_1(k)(2k-4)^2 + C_2(k)(2k-4) + C_3(k)$ is an integer. It forces that $C_j(k)$ $(j = 1, 2, 3)$ must satisfy certains conditions. More precisely, if we let $P(k)$ denote the sum of the main terms, then we have

$$C_1(k)(2k-4)^2 + C_2(k)(2k-4) + C_3(k) = \dim_{\mathbf{C}} S(k; Sp(3, \mathbf{Z})) - P(k),$$

$$C_1(k)(2k+20)^2 + C_2(k)(2k+20) + C_3(k) = \dim_{\mathbf{C}} S(k+12; Sp(3, \mathbf{Z})) - P(k+12),$$

$$C_1(k)(2k+44)^2 + C_2(k)(2k+44) + C_3(k) = \dim_{\mathbf{C}} S(k+24; Sp(3, \mathbf{Z})) - P(k+24).$$

This tells us that $C_j(k)$ $(j = 1, 2, 3)$ can be determined by three consecutive integers $\dim_{\mathbf{C}} S(k; Sp(3, \mathbf{Z}))$, $\dim_{\mathbf{C}} S(k+12; Sp(3, \mathbf{Z}))$, $\dim_{\mathbf{C}} S(k+24; Sp(3, \mathbf{Z}))$ and the sum of the main terms $P(k)$. Now a direct computation with the help of the above observation, we are able to write down the explicit expression of $\dim_{\mathbf{C}} S(k; Sp(3, \mathbf{Z}))$ correctly.

MAIN THEOREM I. *For even integer* $k \geqslant 10$, *the dimension formula for the vector space of Siegel cusp forms of degree three and weight* k *is given by*

$$\dim_{\mathbf{C}} S(k, Sp(3, Z)) = \underline{\text{Sum of Main Terms}} + C_1(k)(2k-4)^2 + C_2(k)(2k-4) + C_3(k)$$

where the main terms and the values of $C_j(k)$ $(j = 1, 2, 3)$ *are given by TABLE I as follows:*

TABLE I Main Terms in the Dimension Formula

No.	Contribution	Conjugacy Classes
1	$2^{-15}3^{-6}5^{-2}7^{-1}(2k-2)(2k-3)(2k-4)^2(2k-5)(2k-6)$	E_6
2	$2^{-15}3^{-4}5^{-1}31(2k-2)(2k-4)^2(2k-6)$	$[1, 1, -1]$
3	$-2^{-13}3^{-3}5^{-1}16(2k-3)(2k-4)(2k-5)$	$E_4 \times \begin{bmatrix} -1 & s \\ 0 & -1 \end{bmatrix}$, $(s \neq 0)$
4	$2^{-10}3^{-5}5^{-1}(2k-3)(2k-4)(2k-5) \times [-2, 0, 2]$	$[1, 1, e^{i\theta}]$, $\theta = \frac{\pi}{3}, \frac{2\pi}{3}, \frac{4\pi}{3}, \frac{5\pi}{3}$ (for Nos. 4–6)
5	$2^{-10}3^{-5}5^{-1}(2k-4)(2k-5) \times [-10, 20, -10]$	
6	$2^{-9}3^{-5}5^{-1}(2k-4) \times [8, 10, -18]$	
7	$-(-1)^{k/2}2^{-12}3^{-2}5^{-1}(2k-4)^2$	$[1, 1, \pm i]$
8	$-2^{-9}3^{-2}5^{-1}(2k-4)$	$[S, E_3]$, rank $S = 2$
9	$2^{-3}3^{-1}5^{-2}(2k-4) \times [1, 0, -1, 3, -3]$	Elements with characteristic polynomial $(X \pm 1)^2(X^4 \pm X^3 + X^2 \pm X + 1)$ (for Nos. 9–10)
10	$2^{-3}3^{-1}5^{-2} \times [-66, 0, 54, -54, 66]$	
11	$\frac{1}{7}$ [1, 0, 1, 0, 0, 0, 0]	Elements of order 7
12	$\frac{1}{9}$ [1, 0, 1, 0, -1, 0, 0, -1, 0]	Elements of order 9
13	$\frac{1}{20}$ [1, 0, 1, 1, -1, -1, 0, -1, -1, 1]	Elements of order 20
14	$\frac{1}{15}$ [1,0,1,0,0,-1,0,0,0,0,0,0,-1,0,0]	Elements of order 30

15 The remaining term is $C_1(k)(2k-4)^2 + C_2(k)(2k-4) + C_3(k)$, where

$$C_1(k) = 2^{-7}3^{-2}[4,2,4,3,3,3] + 2^{-12}3^{-6}[451,1249,451,937,763,937]$$

(TABLE I CONTINUED)

$C_2(k) = -2^{-3}3^{-1} + 2^{-8}3^{-6}[-3010, 783, -4496, -1714, -1161, -1904]$

$C_3(k) = 2^{-4}3^{-6}[5314, 0, 8770, 2560, 2916, 2128]$.

* Here $C(k) = [a_0, a_1, \ldots, a_{m-1}]$ means $C(k) = a_j$ if $k \equiv 2j \pmod{2m}$ for $0 \leqslant j \leqslant m-1$.

MAIN THEOREM II. The dimension formula for the vector space of Siegel cusp forms of degree three with respect to the congruence subgroup $\Gamma_3(2)$ of $\Gamma_3 = Sp(3, \mathbf{Z})$ is given by

$$\begin{aligned}
&\dim_{\mathbf{C}} S(k;\ \Gamma_3(2)) \\
&= [\Gamma_3 : \Gamma_3(2)] \times [2^{-15}3^{-6}5^{-2}7^{-1}(2k-2)(2k-3)(2k-4)^2(2k-5)(2k-6) \\
&\quad + 2^{-15}3^{-4}5^{-1}(2k-2)(2k-4)^2(2k-6) \\
&\quad - 2^{-14}3^{-4}5^{-1}(2k-3)(2k-4)(2k-5) - 2^{-13}3^{-3}(2k-3)(2k-5) \\
&\quad - 2^{-14}3^{-2}5^{-1}(2k-4) + 2^{-13}3^{-1}(2k-4) - 2^{-13}3^{-1} + 2^{-13}3^{-3}]
\end{aligned}$$

for an even integer $k \geqslant 10$, where $[\Gamma_3 : \Gamma_3(2)] = 2^9 3^4 \cdot 35$.

MAIN THEOREM III. The dimension formula for the vector space of Siegel cusp forms of degree three with respect to the principal congruence subgroup $\Gamma_3(N)$ $(N \geqslant 3)$ of $\Gamma_3 = Sp(3, \mathbf{Z})$ is given by

$$\begin{aligned}
&\dim_{\mathbf{C}} S(k;\ \Gamma_3(N)) \\
&= [\Gamma_3 : \Gamma_3(N)] \times [2^{-15}3^{-6}5^{-2}7^{-1}(2k-2)(2k-3)(2k-4)^2(2k-5)(2k-6) \\
&\quad - 2^{-9}3^{-2}5^{-1}(2k-4)N^{-5} + 2^{-7}3^{-3}N^{-6}],
\end{aligned}$$

where k is an even integer greater than 9 and

$$[\Gamma_3 : \Gamma_3(N)] = \frac{1}{2} N^{21} \prod_{p \mid N \ p:\ \text{prime}} (1 - p^{-2})(1 - p^{-4})(1 - p^{-6}).$$

The method we employed here applies to cases of higher degrees. Indeed, we did reduce the problem of finding $\dim_{\mathbf{C}} S(k; Sp(n, Z))$, at least for the case $n = 1, 2, 3$; to the problem of

(1) finding conjugacy classes of $Sp(n, Z)$,

(2) calculating contributions from certain conjugacy classes or families of conjugacy classes

and

(3) determining values of certain constants.

Part of the problem in (1) is treated in [22, 30] . Thus we can write down conjugacy classes of elements whose characteristic polynomials are products of cyclotomic polynomials by an iduction on the degree n . The problem in (2) is treated in [19] in a more general context though not so explicitly. The problem in (3) can be treated by our knowledge of modular forms of lower weight instead of direct computation. In our determination of $\dim_{\mathbf{C}} S(k; Sp(3, Z))$, the constants $C_j(k)$ $(j = 1, 2, 3)$ can be determined uniqued by $\dim_{\mathbf{C}} S(k; Sp(3, Z))$ when $10 \leqslant k \leqslant 44$ and the sum of main terms as shown in TABLE I.

In CHAPTER 1 and 2, we shall determine all conjugacy classes of $Sp(3, Z)$ explicitly for further usage. We began to compute contributions by Theorems in [11] concerning evaluation of integrals involving in Selberg trace formula and conjugacy classes given in CHAPTER 1 and 2. In the final CHAPTER, we shall combine all contributions by the method we mentioned to obtain MAIN THEOREMS in this paper.

This is a continuation of my previous work [11] on the dimension formula of Siegel cusp forms of degree three. I would like to thank my advisor Professor W. L. Baily Jr. at the University of Chicago. Without his constant encouragement, I may give up in the middle owning to the complication of computation.

CHAPTER I

FIXED POINTS AND CONJUGACY CLASSES OF REGULAR ELLIPTIC ELEMENTS IN Sp(3, Z)

1.1. Introduction

In [13] and [14], E. Gottschling studied the fixed points and their isotropy groups of finite order elements in Sp(2, Z). He finally obtained six Sp(2, Z)-inequivalent isolated fixed points as follows:

(1) $Z_1 = \mathrm{diag}\,[i, i]$, (2) $Z_2 = \mathrm{diag}\,[\rho, \rho], \rho = e^{\pi i/3}$,

(2) $Z_3 = \mathrm{diag}\,[i, \rho]$, (4) $Z_4 = \frac{i}{\sqrt{3}}\begin{bmatrix} 2 & 1 \\ 1 & 2 \end{bmatrix}$,

(5) $Z_5 = \begin{bmatrix} \eta & (\eta-1)/2 \\ (\eta-1)/2 & \eta \end{bmatrix}, \quad \eta = \frac{1}{3} + \frac{2\sqrt{2}i}{3}$,

(6) $Z_6 = \begin{bmatrix} \omega & \omega+\omega^{-2} \\ \omega+\omega^{-2} & -\omega^{-1} \end{bmatrix}, \quad \omega = e^{2\pi i/5}$.

The isotropy subgroups at Z_i (i = 1, 2, 3, 4, 5, 6) are groups of order 16, 36, 12, 24, 5 respectively.

By the arguement of [30], these fixed points can be obtained from symplectic embeddings of

$Q(i) \oplus Q(i)$, $Q(\rho) \oplus Q(\rho)$, $Q(i) \oplus Q(\rho)$, $Q(e^{\pi i/6})$,

$Q(e^{\pi i/4})$, $Q(e^{2\pi i/5})$

into $M_4(Q)$. In this CHAPTER, we shall combine the reduction theory of symplectic matrices [5, 6] with the arguements of [22, 30] and obtain all Sp(3, Z)-inequivalent isolated fixed points and conjugacy classes of regular elliptic elements in Sp(3, Z). A table for all representatives and their centralizer in Sp(3, Z)/{±1} of regular elliptic conjugacy classes in Sp(3, Z) is given in 1.4.

1.2 Notations and Basic Results

Let **Z**, **Q**, **R** and **C** denote the ring of integers, the fields of rational, real and complex numbers respectively. The real symplectic matrices of degree n ,

$$Sp(n, \mathbf{R}) = \left\{ M \in M_{2n}(\mathbf{R}) \mid {}^tMJM = J,\ J = \begin{bmatrix} 0 & E_n \\ -E_n & 0 \end{bmatrix} \right\},$$

acts on the generalized half space H_n defined by

$$H_n = \{Z \in M_n(\mathbf{C}) \mid Z = {}^tZ,\quad \operatorname{Im} Z > 0\} .$$

Here $M_{2n}(\mathbf{R})$ is the $2n \times 2n$ matrix ring over **R**, $M_n(\mathbf{C})$ is the $n \times n$ matrix ring over **C**, E_n is the identity of $M_n(\mathbf{C})$ and ${}^tZ$ is the transpose of Z .

A point Z_0 in H_n is called an isolated fixed point of Sp(3, **Z**) if there exists $M = \begin{bmatrix} A & B \\ C & D \end{bmatrix}$ in Sp(3, **Z**) such that Z_0 is the unique solution of the equation,

$$AZ + B = Z(CZ + D),\ Z \in H_n .$$

An element M of $Sp(3, \mathbf{Z})$ is regular elliptic if M has an isolated fixed point [see 10]. Now suppose M is a regular elliptic element of $Sp(3, \mathbf{Z})$, then by the discreteness of $Sp(3, \mathbf{Z})$ and the property that $Sp(3, \mathbf{R})$ acts transitively on H_3, we concluded that

(1) M is an element of finite order,

(2) M is conjugate in $Sp(3, \mathbf{R})$ to $\begin{bmatrix} A & B \\ -B & A \end{bmatrix}$ with $A + Bi = \operatorname{diag}[\lambda_1, \lambda_2, \lambda_3]$, λ_i $(i = 1, 2, 3)$ root of unity and $\lambda_i\lambda_j \neq 1$ for all i, j,

(3) the centralizer of M in $Sp(3, \mathbf{Z})$ is a group of finite order.

By the property (1), we see that the minimal polynomial of M is a product of different cyclotomic polynomials of degree at most 6 as follows:

$$X^2+1,\ X^2-X+1,\ X^2+X+1,\ X^4+1,\ X^4-X^2+1,\ X^4+X^3+X^2+X+1,$$
$$X^4-X^3+X^2-X+1,\ X^6-X^3+1,\ X^6+X^3+1,\ X^6+X^5+X^4+X^3+X^2+X+1,$$
$$X^6-X^5+X^4-X^3+X^2-X+1.$$

For our convenience, we identify $Sp(n_1, \mathbf{R}) \times Sp(n_2, \mathbf{R})$ as a subgroup of $Sp(n_1+n_2, \mathbf{R})$ via the embedding

$$\begin{bmatrix} A & B \\ C & D \end{bmatrix} \times \begin{bmatrix} P & Q \\ R & S \end{bmatrix} \to \begin{bmatrix} A & 0 & B & 0 \\ 0 & P & 0 & Q \\ C & 0 & C & 0 \\ 0 & R & 0 & S \end{bmatrix}$$

Also, we consider the unitary group $U(n)$ as a maximal compact subgroup of $Sp(n, R)$ via the identification

$$A + Bi \to \begin{bmatrix} A & B \\ -B & A \end{bmatrix}.$$

1.3 Reducible Cases.

For each regular elliptic element M in $Sp(3, Z)$, the ring $Q(M)$ is isomorphic to a direct sum of cyclotomic fields which have degree at most 6 since M is a semisimple element. The summand must be equal to one of the following:

$$Q[e^{\pi i/2}],\quad Q[e^{2\pi i/3}],\quad Q[e^{\pi i/4}],\quad Q[e^{2\pi i/5}],$$

$$Q[e^{\pi i/6}],\quad Q[e^{2\pi i/7}],\quad Q[e^{2\pi i/9}] .$$

Now suppose the characteristic polynomial $P(X)$ of M is reducible over $\mathbf{Z}[X]$, then we obtain following 10 possible fixed points for M simply from fixed points of regular elliptic elements of $SL_2(\mathbf{Z})$ and $Sp(2, \mathbf{Z})$.

1. $Z_{01} = \mathrm{diag}[i, i, i]$, 2. $Z_{02} = [\rho, \rho, \rho]$,

3. $Z_{03} = \mathrm{diag}[\rho, i, i]$, 4. $Z_{04} = [i, \rho, \rho]$,

5. $Z_{05} = \begin{bmatrix} i & 0 & 0 \\ 0 & \eta & (\eta-1)/2 \\ 0 & (\eta-1)/2 & \eta \end{bmatrix}$, $\eta = \frac{1}{3} + \frac{2\sqrt{2}i}{3}$,

6. $Z_{06} = \frac{i}{\sqrt{3}} \begin{bmatrix} \sqrt{3} & 0 & 0 \\ 0 & 2 & 1 \\ 0 & 1 & 2 \end{bmatrix}$,

7. $Z_{07} = \begin{bmatrix} i & 0 & 0 \\ 0 & \omega & \omega+\omega^{-2} \\ 0 & \omega+\omega^{-2} & -\omega^{-1} \end{bmatrix}, \quad \omega = e^{2\pi i/5}$,

8. $Z_{08} = \begin{bmatrix} \rho & 0 & 0 \\ 0 & \eta & (\eta-1)/2 \\ 0 & (\eta-1)/2 & \eta \end{bmatrix}$,

9. $Z_{09} = \frac{i}{\sqrt{3}} \begin{bmatrix} 1+\bar{\rho} & 0 & 0 \\ 0 & 2 & 1 \\ 0 & 1 & 2 \end{bmatrix}$,

10. $Z_{10} = \begin{bmatrix} \rho & 0 & 0 \\ 0 & \omega & \omega+\omega^{-2} \\ 0 & \omega+\omega^{-2} & -\omega^{-1} \end{bmatrix}$.

Let G_i (i = 01, 02, 03, 04, 05, 06, 07, 08, 09, 10) be the isotropy group of $Sp(3, Z)/\{\pm 1\}$ at Z_i (i = 01, 02, 03, 04, 05, 06, 07, 08, 09, 10) respectively. Then a direct calculation shows that the order of G_i (i = 01, 02, ..., 10) are 192, 648, 96, 48, 20, 144, 72, 30 respectively. By considering conjugacy classes in G_i (i = 01, 02, ..., 10), we get 72 conjugacy classes of regular elliptic elements of Sp(3, Z) as shown in TABLE II.

REMARK. More precisely, we have the following properties for these isotropy groups. Here $a = \begin{bmatrix} 0 & 1 \\ -1 & 0 \end{bmatrix}$, $b = \begin{bmatrix} 1 & -1 \\ 1 & 0 \end{bmatrix}$, $\alpha = [0, U_1]$, $\beta = [0, U_2]$ with

$$U_1 = \begin{bmatrix} 1 & 0 & 0 \\ 0 & 0 & 1 \\ 0 & 1 & 0 \end{bmatrix} \qquad U_2 = \begin{bmatrix} 0 & 1 & 0 \\ 0 & 0 & 1 \\ 1 & 0 & 0 \end{bmatrix}.$$

(1) G_{01} is a group of order 192 containing the group

$$H_1 = \{M \in G_1 \mid M = a^{\ell} \times z^m \times a^n,\ 0 \leqslant \ell,\ m,n \leqslant 3\}$$

as a normal subgroup of index 6. The coset representatives are $\{E_6, \alpha, \beta, \beta^2, \alpha\beta, \alpha\beta^2\}$ which is isomorphic to the permutation group S_3.

(2) G_{02} is a group of order 648 containing the group

$$H_2 = \{M \in G_2 \mid M = b^{\ell} \times b^m \times b^n,\ 0 \leqslant \ell,\ m,n \leqslant 5\}$$

as a normal subgroup of index 6 with coset representative $\{E_6, \alpha, \beta, \beta^2, \alpha\beta, \alpha\beta^2\}$.

(3) G_{03} is a group of order 96 containing the group

$$H_3 = \{M \in B_3 \mid M = b^{\ell} \times a^m \times a^n,\ 0 \leqslant \ell \leqslant 5,\ 0 \leqslant m,\ n \leqslant 3\}$$

as a normal subgroup of index 2 with coset representatives $\{E, \alpha\}$.

(4) G_{04} is a group of order 144 containing the group

$$H_4 = \{M \in G_4 \mid M = a^{\ell} \times b^m \times b^n,\ 0 \leqslant \ell \leqslant 3,\ 0 \leqslant m,n \leqslant 5\}$$

as a normal subgroup of order 2 with coset representatives $\{E, \alpha\}$.

(5) $G_{05}, G_{06}, G_{07}, G_{08}, G_{09}, G_{10}$ are subgroups of $SL_2(\mathbf{Z}) \times Sp(2, \mathbf{Z})$ of orders 96, 48, 20, 144, 72, 30 respectively. Indeed, they are direct products of

isotropy groups in $SL_2(Z)$ at $z = i$, $z = \rho$ and isotropy groups in $Sp(2, \mathbf{Z})$ at

$$\begin{bmatrix} \eta & (\eta-1)/2 \\ (\eta-1)/2 & \eta \end{bmatrix}, \quad \frac{i}{\sqrt{3}}\begin{bmatrix} 2 & 1 \\ 1 & 2 \end{bmatrix}, \quad \begin{bmatrix} \omega & \omega+\omega^{-2} \\ \omega+\omega^{-2} & -\omega^{-1} \end{bmatrix}.$$

Now we shall show that every regular elliptic clements with reducible characteristic polynomial is conjugate in $Sp(3, \mathbf{Z})$ to one of these 72 conjugacy classes. First we need

LEMMA 1. *Suppose* $M \in Sp(n, Z)$ *with characteristic polynomial* $P(X)$ *satisfying*

(1) $P(X)$ *is a product of two relative prime polynomials* $P_1(X)$, $P_2(X)$ *with integral coefficients of degree* $2n_1$, $2n_2$ $(n_1 + n_2 = n)$ *respectively,*

(2) $P_i(X) = X^{2n_i} P_i(\frac{1}{X})$, $i = 1, 2$,

then there exists $R \in Sp(n, Q)$ *such that* $R^{-1}MR = M_1 \times M_2 \in Sp(n_1, Q) \times Sp(n_2, Q)$. *Furthermore, the characteristic polynomial of* M_1 (resp. M_2) *is* $P_1(X)$ (resp. $P_2(X)$).

Proof. (See LEMMA 1, 2 of [6]).

LEMMA 2. *Let* $M \in Sp(n, \mathbf{Z})$. *Suppose that there exists* $R \in Sp(n, Q)$ *such that*

$$R^{-1}MR = \begin{bmatrix} A & 0 & B & * \\ * & {}^tU & * & * \\ C & 0 & D & * \\ 0 & 0 & 0 & U^{-1} \end{bmatrix},$$

then there exists $R \in Sp(n, \mathbf{Z})$ *such that* $R^{-1}MR$ *has the same form as* $R^{-1}MR$.

Proof. (See Satz 2 of [5]).

THEOREM 1. *Suppose* M *is a regular elliptic element of* $Sp(3, \mathbf{Z})$ *with a reducible characteristic polynomial* $P(X)$, *then* M *is conjugate in* $Sp(3, Z)$ *to an element of* $\bigcup_{i=01}^{10} G_i$.

Proof. Here we only prove three special cases, other cases follows with similar arguements.

(1) $P(X) = (X^2+1)^3$. A representative of M in U(3) is diag[i, i, i]. Thus M is conjugate in $Sp(3, \mathbf{R})$ to $J = \begin{bmatrix} 0 & E \\ -E & 0 \end{bmatrix}$, i.e. there exists $L \in Sp(3, \mathbf{R})$ such that

$$M = L^{-1}JL .$$

With the Iwasawa decomposition of $Sp(3, \mathbf{R})$, we can write

$$L = \begin{bmatrix} A & B \\ -B & A \end{bmatrix} \begin{bmatrix} U & S\,{}^tU^{-1} \\ 0 & {}^tU^{-1} \end{bmatrix}, \quad A+Bi \in U(3) .$$

Since J commutes with $\begin{bmatrix} A & B \\ -B & A \end{bmatrix}$, it follows

$$M = \begin{bmatrix} U & S\,{}^tU^{-1} \\ 0 & {}^tU^{-1} \end{bmatrix}^{-1} J \begin{bmatrix} U & S\,{}^tU^{-1} \\ 0 & {}^tU^{-1} \end{bmatrix} .$$

This forces U, ${}^tU^{-1}$, $S \in GL(3, Z)$. Hence M is conjugate in $Sp(3, \mathbf{Z})$ to $J = \begin{bmatrix} 0 & E \\ -E & 0 \end{bmatrix}$.

(2) $P(X) = (X^2-X+1)^3$. A representative of M in $U(3)$ is $\mathrm{diag}[\rho, \rho, \rho]$ or $\mathrm{diag}[\rho^2, \rho^2, \rho^2]$. On the other hand, $Q(M) \cong Q(\rho)$ as fields and the class number of $Q(\rho)$ is 1 by Theorem 11.1, Chapter 11 of [35]. Hence the number of conjugacy classes of regular elliptic elements with X^2-X+1 as minimal polynomial is 2 by [22] or [30]. Thus M is conjugate in $Sp(3, \mathbf{Z})$ to $\begin{bmatrix} E & -E \\ E & 0 \end{bmatrix}$ or $\begin{bmatrix} E & -E \\ E & 0 \end{bmatrix}^2$.

(3) $P(X) = (X^2+1)(X^4+X^3+X^2+X+1)$. Note that M can be represented in $U(3)$ as $e[1/2, 2/5, 4/5]$ or $e[1/2, 4/5, 8/5]$ or $e[1/2, 6/5, 2/5]$ or $e[1/2, 8/5, 6/5]$. ($e[a, b, c]$ stands for $e^{\pi ia}, e^{\pi ib}, e^{\pi ic}$). In particular, M^5 can be represented in $U(3)$ as $\mathrm{diag}[i, 1, 1]$ or $[-i, 1, 1]$ and has characteristic polynomial $(X^2+1)(X-1)^4$. By Lemma 1 and Lemma 2, there exists $R \in Sp(3, \mathbf{Z})$ such that

$$R^{-1}M^5R = \begin{bmatrix} 0 & 0 & 1 & * \\ * & E_2 & * & * \\ -1 & 0 & 0 & * \\ 0 & 0 & 0 & E_2 \end{bmatrix} \quad \text{or} \quad \begin{bmatrix} 0 & 0 & -1 & * \\ * & E_2 & * & * \\ 1 & 0 & 0 & * \\ 0 & 0 & 0 & E_2 \end{bmatrix}$$

which is conjugate in $Sp(3, \mathbf{Z})$ to

$$\begin{bmatrix} 0 & 1 \\ -1 & 0 \end{bmatrix} \times E_4 \quad \text{or} \quad \begin{bmatrix} 0 & -1 \\ 1 & 0 \end{bmatrix} \times E_4 .$$

Hence we may assume

$$R^{-1}M^5R = \begin{bmatrix} 0 & 1 \\ -1 & 0 \end{bmatrix} \times E_4 \quad \text{or} \quad \begin{bmatrix} 0 & -1 \\ 1 & 0 \end{bmatrix} \times E_4 .$$

Note that the isolated fixed point of $R^{-1}MR$ is contained in the set of fixed points of $R^{-1}M^5R$, i.e. the set

$$Z = \begin{bmatrix} i & 0 & 0 \\ 0 & z_2 & z_{23} \\ 0 & z_{23} & z_3 \end{bmatrix}, \quad \text{Im } Z > 0 .$$

Now it is easy to see that the isolated fixed point of $R^{-1}MR$ is $SL_2(Z) \times Sp(2, Z)$-equivalent to Z_{07} and M is conjugate in $Sp(3, \mathbf{Z})$ to an element of G_{07}.

In the remaining sections, we shall determined conjugacy classes of regular elliptic elements of order 9 and 7.

1.4 Symplectic embeddings of $Q(e^{2\pi i/9})$ and $Q(e^{2\pi i/7})$.

For our converience, we denote $e^{2\pi i/9}$ by ζ. Note that $Q(\zeta)$ is the splitting field of the cyclotomic polynomial X^6+X^3+1 and contains the total real number field $Q(\zeta+\zeta^{-1})$ which is the splitting field of X^3-3X+1. By a symplectic embedding of $Q(\zeta)$ into $M_6(Q)$, we mean a injection from $Q(\zeta)$ into $M_6(Q)$ such that ζ is mapped into a symplectic matrix M and $Q(\zeta)$ $Q(M)$ as fields [30].

LEMMA 3. *Let* M *be an element of* $Sp(3, \mathbf{Z})$ *of order* 9. *Then* M *is conjugate in* $Sp(3, \mathbf{R})$ *to one of the following*:

$$[\zeta, \zeta^4, \zeta^7], \quad [\zeta, \zeta^2, \zeta^4], \quad [\zeta, \zeta^2, \zeta^5], \quad [\zeta, \zeta^5, \zeta^7],$$

$$[\zeta^2, \zeta^4, \zeta^8], \quad [\zeta^2, \zeta^5, \zeta^8], \quad [\zeta^4, \zeta^7, \zeta^8], \quad [\zeta^5, \zeta^7, \zeta^8].$$

Proof. The minimal polynomial of M is X^6+X^3+1 which can be factored into

$$(X-\zeta)(X-\zeta^2)(X-\zeta^4)(X-\zeta^5)(X-\zeta^7)(X-\zeta^8)$$

$$= [X^2-(\zeta+\zeta^{-1})X+1][X^2-(\zeta^2+\zeta^{-2})X+1][X^2-(\zeta^4+\zeta^{-4})X+1].$$

Hence M is conjugate in $Sp(3, \mathbf{R})$ to

$$\begin{bmatrix} \cos\theta & \pm\sin\theta \\ \pm\sin\theta & \cos\theta \end{bmatrix} \times \begin{bmatrix} \cos 2\theta & \pm\sin 2\theta \\ \pm\sin 2\theta & \cos 2\theta \end{bmatrix} \times \begin{bmatrix} \cos 4\theta & \pm\sin 4\theta \\ \pm\sin 4\theta & \cos 4\theta \end{bmatrix}, \quad \theta = \frac{2\pi}{9}.$$

Note that the above eight elements of $Sp(3, \mathbf{R})$ are represented by the prescribled elements in $U(3)$ as in our Lemma.

LEMMA 4. *The number of conjugacy classes of regular elliptic elements of order* 9 *in* $Sp(3, Z)$ *is* 8.

Proof. The ideal class number of $Q(\zeta)$ is 1 as given in Theorem 11.1 of [35], hence the number of conjugacy classes of regular elliptic elements of order 9 is given by

$$[E_0 : N(E)],$$

where

E: the group of units in $Q(\zeta)$,

E_0: the group of units in $Q(\zeta+\zeta^{-1})$,

$N(E) = \{u\bar{u} \mid u \in E\}$,

according to the arguement of [22] or [30].

The group of units for cyclotomic fields are determined in Chap. 8 of [35]. Applying to our case, we get $[E_0 : N(E)] = 8$ when the cyclotomic field is $Q(\zeta)$.

There are two conjugacy classes of elements of order 9 appeared in the isotropy group G_{02} of $Z_{02} = \mathrm{diag}[\rho, \rho, \rho]$. Indeed, if we let $M = \begin{bmatrix} A & B \\ C & D \end{bmatrix}$ with

$$A = \begin{bmatrix} 0 & 0 & 0 \\ 1 & 0 & 1 \\ 0 & 1 & 0 \end{bmatrix}, \qquad B = \begin{bmatrix} 0 & 0 & -1 \\ 0 & 0 & 0 \\ 0 & 0 & 0 \end{bmatrix},$$

$$C = \begin{bmatrix} 0 & 0 & 1 \\ 0 & 0 & 0 \\ 0 & 0 & 0 \end{bmatrix}, \quad \text{and} \quad D = \begin{bmatrix} 0 & 0 & -1 \\ 1 & 0 & 0 \\ 0 & 0 & 0 \end{bmatrix},$$

then it is a direct verification to shows that

(1) M is an element of order 9 .

(2) M can be represented in $U(3)$ as $[\zeta, \zeta^4, \zeta^7]$ or $[\zeta^2, \zeta^8, \zeta^5]$.

(3) $M^3 = \begin{bmatrix} 0 & -E \\ E & -E \end{bmatrix}$ has an isolated fixed point at Z_{02} .

Now we begin to look for other 6 conjugacy classes of regular elliptic elements of order 9 in Sp(3, Z) .

THEOREM 2. *Suppose* α, β, γ *are distinct roots of the equation* $X^3-3X+1 = 0$ (*or more precisely*, $\alpha = 2\cos\frac{2\pi}{9}$, $\beta = 2\cos\frac{4\pi}{9}$, $\gamma = 2\cos\frac{8\pi}{9}$),

$$A = \begin{bmatrix} 0 & 1 & 0 \\ 1 & -1 & 1 \\ 0 & 1 & 1 \end{bmatrix}, \quad \Omega = \frac{1}{3}\begin{bmatrix} -3+\alpha+\alpha^2 & -3+\alpha+\alpha^2 & -3+\alpha+\alpha^2 \\ -1+\alpha^2 & -1+\alpha^2 & -1+\alpha^2 \\ 1+\alpha & 1+\alpha & 1+\alpha \end{bmatrix},$$

and $M = \begin{bmatrix} A & E \\ -E & 0 \end{bmatrix}$, *then*

(1) M *is an element of order* 9 *in* Sp(3, **Z**) *and has an isolated fixed point at*

$$Z_{11} = -\frac{1}{2}A + i\Omega(E - \frac{1}{4}\,{}^t\Omega A^2\Omega)^{\frac{1}{2}}\,{}^t\Omega,$$

(2) M is conjugate in Sp(3, **R**) to $[\zeta, \zeta^2, \zeta^4]$ of U(3) ,

(3) the centralizer of M in Sp(3, **Z**)/{±1} is a group of order 9 .

Proof. (1) Since the characteristic polynomial of M is X^6+X^3+1 , it follows that M is an element of order 9 in Sp(3, Z) . Note that $\frac{1}{3}\,{}^t[-3+\alpha+\alpha^2, -1+\alpha^2, 1+\alpha]$ is the normalized eigenvector of A corresponding to the eigenvalue α , it follows

$$^t\Omega A\Omega = \mathrm{diag}[\alpha, \beta, \gamma]$$

and

$$(E - \tfrac{1}{4}\, {}^t\Omega A\Omega)^{\frac{1}{2}} = \mathrm{diag}[(1-\alpha^2/4)^{\frac{1}{2}}, (1-\beta^2/4)^{\frac{1}{2}}, (1-\gamma^2/4)^{\frac{1}{2}}] .$$

Now it is a direct verification to show that $AZ_{11} = Z_{11}A$ and $Z_{11}^2 + AZ_{11} + E = 0$. Thus $Z_{11} = -\frac{1}{2}A + i\Omega(E - \frac{1}{4}\, {}^t\Omega A^2\Omega)^{\frac{1}{2}}\, {}^t\Omega$ is a fixed point of M. But M has exactly one fixed point by Lemma 3, hence Z_{11} is the unique isolated fixed point of M.

(2) Let $R = \begin{bmatrix} \Omega & 0 \\ 0 & \Omega \end{bmatrix}$. Then $R \in Sp(3, \mathbf{R})$ and

$$R^{-1}MR = \begin{bmatrix} \alpha & 1 \\ -1 & 0 \end{bmatrix} \times \begin{bmatrix} \beta & 1 \\ -1 & 0 \end{bmatrix} \times \begin{bmatrix} \gamma & 1 \\ -1 & 0 \end{bmatrix} .$$

Note that $R^{-1}MR$ is conjugate in $Sp(3, \mathbf{R})$ to

$$\begin{bmatrix} \cos\theta & \sin\theta \\ -\sin\theta & \cos\theta \end{bmatrix} \times \begin{bmatrix} \cos 2\theta & \sin 2\theta \\ -\sin 2\theta & \cos 2\theta \end{bmatrix} \times \begin{bmatrix} \cos 3\theta & \sin 3\theta \\ -\sin 3\theta & \cos 3\theta \end{bmatrix}, \quad \theta = \frac{2\pi}{9}$$

because $\begin{bmatrix} 2\cos\theta & 1 \\ -1 & 0 \end{bmatrix}$ is conjugate in $SL_2(\mathbf{R})$ to $\begin{bmatrix} \cos\theta & \sin\theta \\ -\sin\theta & \cos\theta \end{bmatrix}$. This proves our assertion in (2).

(3) Let $\mathbf{C}(M, \mathbf{Z})$ be the centralizer of M in $Sp(3, \mathbf{Z})/\{\pm 1\}$. Suppose γ is an element of $\mathbf{C}(M, \mathbf{Z})$, then

$$M(\gamma(Z_{11})) = \gamma(M(Z_{11})) = \gamma(Z_{11}) .$$

Since Z_{11} is the only fixed point of M, this forces $\gamma(Z_{11}) = Z_{11}$.

Note that ${}^t\Omega Z_{11}\Omega = R(Z_{11}) = \mathrm{diag}[-\bar{\zeta}, -\bar{\zeta}^2, -\bar{\zeta}^4]$. Here $R = \begin{bmatrix} \Omega & 0 \\ 0 & \Omega \end{bmatrix}$ as in (2). From $\gamma(Z_{11}) = Z_{11}$, we get

$$R\gamma R^{-1}({}^t\Omega Z_{11}\Omega) = {}^t\Omega Z_{11}\Omega .$$

It follows

$$R\gamma R^{-1} = \begin{bmatrix} a & b \\ c & d \end{bmatrix} \times \begin{bmatrix} a' & b' \\ c' & d' \end{bmatrix} \times \begin{bmatrix} a'' & b'' \\ c'' & d'' \end{bmatrix}$$

with

$$\begin{cases} -a\bar{\zeta} + b = c\bar{\zeta}^2 - d\bar{\zeta}, \quad ad - bc = 1, \\ -a'\bar{\zeta}^2 + b' = c'\bar{\zeta}^4 - d'\bar{\zeta}^2, \quad a'd' - b'c' = 1, \\ -a''\bar{\zeta}^4 + b'' = c''\bar{\zeta}^8 - d''\bar{\zeta}^4, \quad a''d'' - b''c'' = 1. \end{cases}$$

The general solution of a, b, c, d is given by

$$\begin{cases} a = \cos\theta - \cot\frac{2\pi}{9}\sin\theta, \quad b = -\sec\frac{2\pi}{9}\sin\theta, \\ c = \sec\frac{2\pi}{9}\sin\theta, \quad d = \cos\theta + \cot\frac{2\pi}{9}\sin\theta, \end{cases} \qquad \theta \in R .$$

The characteristic polynomial of $\begin{bmatrix} a & b \\ c & d \end{bmatrix}$ is $X^2 - 2\cos\theta X + 1$, hence $2\cos\theta$ is an algebraic integer of degree 1 or 3. On the other hand, the fact that γ is an element of finite order implies $e^{i\theta}$ is a root of unity. Now we have the following cases:

Case I. If $2\cos\theta$ is an algebraic integer of degree 3, then the characteristic polynomial of γ is an irreducibel polynomial of degree 6. Since M satisfies this case, so that the characteristic polynomial of γ is X^6+X^3+1 or X^6-X^3+1. This leads to the fact that $\theta = \frac{2\pi}{9}$ or $\frac{4\pi}{9}$ or $\frac{8\pi}{9}$ and γ is one of the following elements:

$$\pm M,\ \pm M^2,\ \pm M^5,\ \pm M^7,\ \pm M^8 .$$

Case II. If $2\cos\theta = 1$, then γ is an element of order 3. Then $\gamma = \pm M^3$ or $\pm M^6$ by a direct calculation.

Case III. If $2\cos\theta = 2$, then $\gamma = E_6$.

Case IV. If $2\cos\theta = 0$, then

$$\gamma = R^{-1}\left\{\begin{bmatrix} -\cot\eta & -\sec\eta \\ \sec\eta & \cot\eta \end{bmatrix} \times \begin{bmatrix} -\cot 2\eta & -\sec 2\eta \\ \sec 2\eta & \cot 2\eta \end{bmatrix} \times \begin{bmatrix} -\cot 4\eta & -\sec 4\eta \\ \sec 4\eta & \cot 4\eta \end{bmatrix}\right\} R$$

with $\eta = \frac{2\pi}{9}$. Such a γ is not an integral matrix.

By the above discussion, we conclude that $C(M, \mathbf{Z})$ is a group of order 9 generated by M.

With the same argument, we get the following result simply replace the role of $e^{2\pi i/9}$ by $e^{2\pi i/7}$.

LEMMA 5. Let M be an element of $Sp(3, \mathbf{Z})$ or order 7. Then M is conjugate in $Sp(3, \mathbf{R})$ to one of the following: $(v = e^{2\pi i/7})$

$$[v, v^2, v^3], \quad [v, v^2, v^4], \quad [v, v^4, v^5], \quad [v, v^3, v^5 ,$$

$$[v^2, v^3, v^4], [v^2, v^4, v^6], \quad [v^4, v^5, v^6], [v^3, v^5, v^6] .$$

LEMMA 6. *The number of conjugacy classes of regular elliptic elements of order* 7 *in* $Sp(3, \mathbf{Z})$ *is* 8 .

THEOREM 3. *Suppose* α, β, γ *are distinct roots of the equation* X^3+X^2-2X+1 , *(or more precisely,* $\alpha = 2\cos\frac{2\pi}{7}$, $\beta = 2\cos\frac{4\pi}{7}$, $\gamma = 2\cos\frac{6\pi}{7}$),

$$B = \begin{bmatrix} 0 & 1 & 0 \\ 1 & 0 & 1 \\ 0 & 1 & -1 \end{bmatrix} \qquad \Omega' = \begin{bmatrix} \frac{\alpha+\alpha^2}{1+3\alpha} & \frac{\beta+\beta^2}{1+3\beta} & \frac{\gamma+\gamma^2}{1+3\gamma} \\ \frac{1+2\alpha}{1+3\alpha} & \frac{1+2\beta}{1+3\beta} & \frac{1+2\gamma}{1+3\gamma} \\ \frac{\alpha^2}{1+3\alpha} & \frac{\beta^2}{1+3\beta} & \frac{\gamma^2}{1+3\gamma} \end{bmatrix}$$

and $M = \begin{bmatrix} B & E \\ -E & 0 \end{bmatrix}$, *then*

(1) M *is an element of order* 7 *in* $Sp(3, \mathbf{Z})$ *and has an isolated fixed point at*

$$Z_{12} = -\frac{1}{2}B + i\Omega'(E - \frac{1}{4}\,{}^t\Omega' B^2 \Omega')^{\frac{1}{2}}\,{}^t\Omega' ,$$

(2) M *is conjugate in* $Sp(3, \mathbf{R})$ to $[v, v^2, v^3]$,

(3) *the centralizer of* M *in* $Sp(3, \mathbf{Z})/\{\pm 1\}$ *is a group of order* 7 *generated by* M .

Note that $[v, v^2, v^4]$ and $[v^3, v^6, v^5]$ are exclusive in the set of all powers of $[v, v^2, v^3]$. To find all representatives for elliptic conjugacy classes of order 7, it suffices to get a representative which is conjugate in $Sp(3, \mathbf{R})$ to $[v, v^2, v^4]$.

THEOREM 4. *Let* B, Ω' *be matrices as in Theorem 3,* $U = \mathrm{diag}[1, 1, -1]$ *and* $M = \begin{bmatrix} B & E+B \\ -(E+B)^{-1} & 0 \end{bmatrix}$. *Then*

(1) M *is an element of order* 7 *in* $Sp(3, Z)$ *with isolated fixed point at*

$$Z_{13} = -\frac{1}{2}B(B+E) + i\Omega'[(E - \frac{1}{4}{}^t\Omega' B^2 \Omega')^{\frac{1}{2}}{}^t\,'(B+E)U\,{}^t\Omega'],$$

(2) M *is conjugate in* $Sp(3, \mathbf{R})$ *to* $[v, v^2, v^4]$,

(3) *the centralizer of* M *in* $Sp(3, \mathbf{Z})/\{\pm 1\}$ *is a finite group of order* 7 *generated by* M.

Proof. Since $\det(E+B) = -1$ and

$${}^t\Omega'(E+B)\Omega' = \mathrm{diag}[1 + 2\cos\frac{2\pi}{7},\ 1 + 2\cos\frac{4\pi}{7},\ 1 + 2\cos\frac{6\pi}{7}]$$

has signature $+, +, -$, it follows $M \in Sp(3, \mathbf{Z})$ and M is conjugate in $Sp(3, \mathbf{R})$ to

$$M' = \begin{bmatrix} 2\cos\theta & 1 \\ -1 & 0 \end{bmatrix} \times \begin{bmatrix} 2\cos 2\theta & 1 \\ -1 & 0 \end{bmatrix} \times \begin{bmatrix} 2\cos 3\theta & -1 \\ 1 & 0 \end{bmatrix}, \quad \theta = \frac{2\pi}{7}$$

Indeed, if we let

$R' = \Omega'\Lambda$ with $\Lambda = \mathrm{diag}[(1+2\cos\frac{2\pi}{7})^{-\frac{1}{2}}, (1+2\cos\frac{2\pi}{7})^{-\frac{1}{2}}, (-1-2\cos\frac{2\pi}{7})^{-\frac{1}{2}}]$

then $(R')^{-1}MR' = M'$. Hence (1) and (2) follow as a direct calculation. By a similar arguement as (3) of Theorem 2, we get (2).

By Theorem 1, Theorem 2, Theorem 3 and Theorem 4, we obtain the following table for conjugacy classes of regular elliptic elements in $Sp(3, Z)$.

TABLE II: Regular Elliptic Cinjugacy Classes of $Sp(3, \mathbf{Z})$

Here e[a,b,c] stands for diag $e[^{\pi ia}, e^{\pi ib}, e^{\pi ic}]$.

No	Representative in U(3)	Minimal Polynomial	order of Centralizer	No. of Conjugates in Isotropy group
1	e[1/2, 1/2, 1/2]	X^2+1	192	1
2	e[1/2, 1/4, 5/4]	$(X^2+1)(X^4+1)$	16	12
3	e[1/2, 3/4, 7/4]	$(X^2+1)(X^4+1)$	16	12
4	e[1/6, 5/6, 9/6]	X^6+1	6	32
5	e[1/3, 1/3, 1/3]	X^2-X+1	648	1
6	e[2/3, 1/3, 1/3]	X^4+X^2+1	216	3
7	e[4/3, 1/3. 1/3]	X^4+X^2+1	216	3
8	e[2/3, 2/3, 1/3]	X^4+X^2+1	216	3
9	e[2/3, 2/3, 2/3]	X^2+X+1	648	1
10	e[5/3, 2/3, 2/3]	X^4+X^2+1	216	3
11	e[1/3, 1/3, 4/3]	X^4+X^2+1	36	18
12	e[2/3, 2/3, 5/3]	X^4+X^2+1	36	18
13	e[1/3, 1/6, 7/6]	X^4+X^2+1	36	18
14	e[2/3, 1/6, 7/6]	X^4+X^2+1	36	18

(TABLE II CONTINUED)

15	e[1/3, 5/6, 11/6]	X^4+X^2+1	36	18
16	e[2/3, 5/6, 11/6]	X^4+X^2+1	36	18
17	e[2/9, 8/9, 14/9]	X^6+X^3+1	9	72
18	e[4/9, 10/9, 16/9]	X^6+X^3+1	9	72
19	e[1/3, 1/2, 1/2]	$(X^2-X+1)(X^2+1)$	96	1
20	e[1/3, 1/2, 1/2]	$(X^2-X+1)(X^2+1)$	96	1
21	e[4/3, 1/2, 1/2]	$(X^2+X+1)(X^2+1)$	96	1
22	e[5/3, 1/2, 1/2]	$(X^2-X+1)(X^2+1)$	96	1
23	e[1/3, 1/4, 5/4]	$(X^2-X+1)(X^4+1)$	24	4
24	e[2/3, 1/4, 5/4]	$(X^2+X+1)(X^4+1)$	24	4
25	e[1/3, 3/4, 7/4]	$(X^2-X+1)(X^4+1)$	24	4
26	e[2/3, 3/4, 7/4]	$(X^2+X+1)(X^4+1)$	24	4
27	e[1/2, 1/3, 1/3]	$(X^2+1)(X^2-X+1)$	144	1
28	e[3/2, 1/3, 1/3]	$(X^2+1)(X^2-X+1)$	144	1
29	e[1/2, 2/3, 2/3]	$(X^2+1)(X^2-X+1)$	144	1
30	e[3/2, 2/3, 2/3]	$(X^2+1)(X^2+X+1)$	144	1
31	e[1/2, 2/3, 1/3]	$(X^2+1)(X^4+X^2+1)$	72	2
32	e[3/2, 2/3, 1/3]	$(X^2+1)(X^4+X^2+1)$	72	2
33	e[1/2, 4/3, 1/3]	$(X^2+1)(X^4+X^2+1)$	72	2
34	e[1/2, 5/3, 2/3]	$(X^2+1)(X^4+X^2+1)$	72	2
35	e[1/2, 1/6, 7/6]	$(X^2+1)(X^4-X^2+1)$	24	6
36	e[1/2, 1/3, 4/3]	$(X^2+1)(X^4+X^2+1)$	24	6
37	e[1/2, 5/6, 11/6]	$(X^2+1)(X^4-X^2+1)$	24	6
38	e[1/2, 2/3, 5/3]	$(X^2+1)(X^4+X^2+1)$	24	6
39	e[1/2, 1/4, 3/4]	$(X^2+1)(X^4+X^2+1)$	16	6
40	e[1/2, 5/4, 7/4]	$(X^2+1)(X^4+X^2+1)$	16	6
41	e[1/2, 1/3, 2/3]	$(X^2+1)(X^4+X^2+1)$	24	2
42	e[3/2, 1/3, 2/3]	$(X^2+1)(X^4-X^2+1)$	24	2

(TABLE II CONTINUED)

43	e[1/2, 2/5, 4/5]	$(X^2+1)P_1(X)$	20	1
44	e[1/2, 4/5, 8/5]	$(X^2+1)P_1(X)$	20	1
45	e[1/2, 6/5, 2/5]	$(X^2+1)P_1(X)$	20	1
46	e[1/2, 8/5, 6/5]	$(X^2+1)P_1(X)$	20	1
47	e[3/2, 2/5, 4/5]	$(X^2+1)P_1(-X)$	20	1
48	e[3/2, 4/5, 8/5]	$(X^2+1)P_1(-X)$	20	1
49	e[3/2, 6/5, 2/5]	$(X^2+1)P_1(-X)$	20	1
50	e[3/2, 8/5, 6/5]	$(X^2+1)P_1(-X)$	20	1
51	e[1/3, 1/4, 3/4]	$(X^2-X+1)(X^4+1)$	24	6
52	e[2/3, 1/4, 3/4]	$(X^2-X+1)(X^4+1)$	24	6
53	e[4/3, 1/4, 3/4]	$(X^2+X+1)(X^4+1)$	24	6
54	e[5/3, 1/4, 3/4]	$(X^2+X+1)(X^4+1)$	24	6
55	e[1/3, 1/3, 2/3]	X^4+X^2+1	36	4
56	e[2/3, 1/3, 2/3]	X^4+X^2+1	36	4
57	e[1/3, 2/5, 4/5]	$(X^2-X+1)P_1(X)$	30	1
58	e[2/3, 2/5, 4/5]	$(X^2-X+1)P_1(X)$	30	1
59	e[4/3, 2/5, 4/5]	$(X^2+X+1)P_1(X)$	30	1
60	e[5/3, 2/5, 4/5]	$(X^2+X+1)P_1(X)$	30	1
61	e[1/3, 4/5, 8/5]	$(X^2-X+1)P_1(X)$	30	1
62	e[2/3, 4/5, 8/5]	$(X^2-X+1)P_1(X)$	30	1
63	e[4/3, 4/5, 8/5]	$(X^2+X+1)P_1(X)$	30	1
64	e[5/3, 4/5, 8/5]	$(X^2+X+1)P_1(X)$	30	1
65	e[1/3, 6/5, 2/5]	$(X^2-X+1)P_1(X)$	30	1
66	e[2/3, 6/5, 2/5]	$(X^2-X+1)P_1(X)$	30	1
67	e[4/3, 6/5, 2/5]	$(X^2+X+1)P_1(X)$	30	1
68	e[5/3, 6/5, 2/5]	$(X^2+X+1)P_1(X)$	30	1
69	e[1/3, 8/5, 6/5]	$(X^2-X+1)P_1(X)$	30	1

* Here $P_1(X) = X^4+X^3+X^2+X+1$ and $P_2(X) = X^6+X^5+X^4+X^3+X^2+X+1$.

(TABLE II CONTINUED)

70	e[2/3, 8/5, 6/5]	$(X^2-X+1)P_1(X)$	30	1
71	e[4/3, 8/5, 6/5]	$(X^2+X+1)P_1(X)$	30	1
72	e[5/3, 8/5, 6/5]	$(X^2+X+1)P_1(X)$	30	1
73	e[2/9, 4/9, 8/9]	X^6+X^3+1	9	1
74	e[4/9, 8/9, 16/9]	X^6+X^3+1	9	1
75	e[8/9,16/9, 14/9]	X^6+X^3+1	9	1
76	e[10/9, 2/9, 4/9]	X^6+X^3+1	9	1
77	e[14/9,10/9, 2/9]	X^6+X^3+1	9	1
78	e[16/9,14/9,10/9]	X^6+X^3+1	9	1
79	e[2/7, 4/7, 6/7]	$P_2(X)$	7	1
80	e[4/7, 8/7, 12/7]	$P_2(X)$	7	1
81	e[6/7,12/7, 4/7]	$P_2(X)$	7	1
82	e[8/7, 2/7, 10/7]	$P_2(X)$	7	1
83	e[10/7, 6/7, 2/7]	$P_2(X)$	7	1
84	e[12/7,10/7, 8/7]	$P_2(X)$	7	1
85	e[2/7, 4/7, 8/7]	$P_2(X)$	7	3
86	e[6/7, 12/7,10/7]	$P_2(X)$	7	3

1.5 Application

Contributions from conjugacy classes of regular elliptic elements in Sp(n, Z) to the dimension formula for Siegel cusp forms of degree n and weight k [10] is given by

$$\Sigma |C(M, Z)|^{-1} \prod_{i=1}^{n} \bar{\lambda}_i^k \prod_{i \leqslant j} (1 - \bar{\lambda}_i \bar{\lambda}_j)^{-1} .$$

Here the summation if M ranges over all conjugacy classes of regular elliptic elements in Sp(n, **Z**). M is conjugate in

$Sp(n, \mathbf{R})$ to $\begin{bmatrix} A & B \\ -B & A \end{bmatrix}$ with $A+Bi = \mathrm{diag}[\lambda_1, \lambda_2, \ldots, \lambda_n]$, $\lambda_i\lambda_j \neq 1$ for all i,j and $C(M, \mathbf{Z})$ is the centralizer of M in $Sp(3, \mathbf{Z})$. Applying this formula to the case $n = 3$, we get all contributions from 86 regular elliptic conjugacy classes in $Sp(3, \mathbf{Z})$.

For the case $n = 1$ and $n = 2$, contribution from a particular regular elliptic conjugacy class appeared to be a residue of a generating function at a simple pole. For example, the contribution from conjugacy class of regular elliptic element of order 5 in $Sp(2, \mathbf{Z})$ is given by

$$K = \frac{1}{25}[\bar{\omega}^{6k}(1-\bar{\omega}^{2}) + \bar{\omega}^{2k}(1-\bar{\omega}^{4}) + \bar{\omega}^{8k}(1-\bar{\omega}^{6}) + \bar{\omega}^{4k}(1-\bar{\omega}^{8})] ,$$

$$\omega = e^{\pi i/5} ,$$

which is precisely the negative of the sum of residues of the function

$$\frac{1}{(1-T^4)(1-T^6)(1-T^{10})(1-T^{12})T^{k+1}}$$

at $T = e^{i\theta}$ with $\theta = \pm\pi/5, \pm 2\pi/5, \pm 3\pi/5, \pm 4\pi/5$ when k is even.

It is easy to see that the total contribution from conjugacy classes of elements of order 2 or 3 in $SL_2(Z)$ is the negative of the sum of residues of the function

$$\frac{1}{(1-T^4)(1-T^6)T^{k+1}}$$

at $T = e^{i\theta}$ with $\theta = \pm\pi/2, \pm\pi/3, \pm 2\pi/3$ when k is even.

Note that $\dfrac{1}{(1-T^4)(1-T^6)}$ and $\dfrac{1}{(1-T^4)(1-T^6)(1-T^{10})(1-T^{12})}$ are well-known to be generating functions of dimension formulas for modular forms of degree 1 and degree 2 respectively. It is hopeful to find a generating function of dimension formular for modular forms of degree three by computing contributions from conjugacy classes of regular elliptic elements in $Sp(3, Z)$.

CHAPTER II

CONJUGACY CLASSES OF THE MODULAR GROUP $Sp(3, \mathbf{Z})$

2.1 Introduction

Conjugacy classes of regular elliptic elements in $Sp(3, \mathbf{Z})$ are completely determined in the previous CHAPTER by considering conjugacy classes of 13 finite isotropy groups in $Sp(3, Z)$ at 13 $Sp(3, \mathbf{Z})$-inequivalent isolated fixed points. If we exclude conjugacy classes with zero contributions [11], we find that conjugacy classes of remaining elements in $Sp(3, \mathbf{Z})$ have representatives in Γ_3^2, Γ_3^1 and Γ_3^0, the stabilizer in $Sp(3, \mathbf{Z})$ of three dimensional cusp, one dimensional cusp and zero dimensional cusp respectively. We shall obtain a complete list of conjugacy classes (except those with zero contributions) of the full modular group $Sp(3, \mathbf{Z})$ for further application by extracting representatives from $\Gamma_3^2 \cup \Gamma_3^1 \cup \Gamma_3^0$.

2.2 Basic results

Let $\Gamma_3 = Sp(3, \mathbf{Z})$ be the modular group of degree 3, specifically,

$$Sp(3, \mathbf{Z}) = \left\{ M \in M_6(Z) \mid {}^tMJM = J,\ J = \begin{bmatrix} 0 & E_3 \\ -E_3 & 0 \end{bmatrix} \right\}.$$

If we write $M = \begin{bmatrix} A & B \\ C & D \end{bmatrix}$, then necessary and sufficient conditions for $M \in Sp(3, Z)$ are

(1) $A^tD - B^tC = E_3$,

(2) $A^tB = B^tA$, $C^tD = D^tC$.

Here E_3 is the identity of the matrix ring $M_3(Z)$ and ${}^tL$ is the transpose of L .

To compute the dimension of Siegel cusp forms of degree three with respect to $Sp(3, \mathbf{Z})$ via Selberg trace formula (as in [8], [11], [16]) or via Riemann-Roch-Hirzebruch theorem (as in [31], [34]), one first has to classify conjugacy classes of $Sp(3, \mathbf{Z})$ or, at least, classify those conjugacy classes of elements whose characteristic polynomials are products of cyclotomic polynomials by excluding conjugacy classes with zero contributions aas shown in [11].

Indeed, we have the following theorem after a long computation.

THEOREM 1. ([11], Chapter III). *Suppose* $M \in Sp(3, \mathbf{Z})$ *and the conjugacy class represented by* M *has a possible nonzero contribution to the dimension formula, then* M *is conjugate in* $Sp(3, \mathbf{R})$ *to one of the following:*

(1) $\begin{bmatrix} A & B \\ -B & A \end{bmatrix}$, $A + Bi = \mathrm{diag}[\lambda_1, \lambda_2, \lambda_3]$, $\lambda_1, \lambda_2, \lambda_3$ *are roots of unity and* $\lambda_i\lambda_j \neq 1$ *for all* $1 \leqslant i \leqslant j \leqslant 3$;

(2) $\begin{bmatrix} A' & B' \\ -B' & A' \end{bmatrix} \times \begin{bmatrix} 1 & s \\ 0 & 1 \end{bmatrix}$, $A' + B'i = \mathrm{diag}[\lambda_1, \lambda_2]$, λ_1, λ_2 *are roots of unity and* $\lambda_1^2, \lambda_1\lambda_2, \lambda_2^2 \neq 1$,

(3) $\begin{bmatrix} a & b \\ -b & a \end{bmatrix} \times [S, U]$, $a^2 + b^2 = 1$, $U = \begin{bmatrix} 1 & 0 \\ 0 & -1 \end{bmatrix}$ <u>or</u>

$$\begin{bmatrix} \cos\theta & \sin\theta \\ -\sin\theta & \cos\theta \end{bmatrix}.$$

(4) $\begin{bmatrix} P & Q \\ -Q & P \end{bmatrix} \begin{bmatrix} E & S \\ 0 & E \end{bmatrix}$, $P + Qi = \begin{bmatrix} a+bi & 0 & 0 \\ 0 & a & b \\ 0 & -b & a \end{bmatrix}$,

$$S = \begin{bmatrix} 0 & s & 0 \\ s & s' & 0 \\ 0 & 0 & s' \end{bmatrix}, \quad (b \neq 0),$$

(5) $[S, U]$ <u>with</u> $U = [1, 1, -1]$ <u>or</u>

$$U = \begin{bmatrix} 1 & 0 & 0 \\ 0 & \cos\theta & \sin\theta \\ 0 & -\sin\theta & \cos\theta \end{bmatrix}.$$

Elements in (1) are regular elliptic elements (elements with a unique fixed point) and their conjugacy classes have been completely determined in the previous CHAPTER. Now it suffices to find conjugacy classes of elements with property (2) or (4) or (5).

Let Γ_3^2, Γ_3^1 and Γ_3^0 be subgroups of Γ_3 defined as follows:

$$\Gamma_3^2 = \left\{ M \in Sp(3, Z) \mid M = \begin{bmatrix} A & 0 & B & * \\ * & 1 & * & * \\ C & 0 & D & * \\ 0 & 0 & 0 & 1 \end{bmatrix}, \begin{bmatrix} A & B \\ C & D \end{bmatrix} \in Sp(2, \mathbf{Z}) \right\}$$

$$\Gamma_3^1 = \left\{ M \in Sp(3, \mathbf{Z}) \middle| M = \begin{bmatrix} a & 0 & b & * \\ * & U & * & * \\ c & 0 & d & * \\ 0 & 0 & 0 & {}^tU^{-1} \end{bmatrix}, \begin{bmatrix} a & b \\ c & d \end{bmatrix} \in SL_2(Z) \right\},$$

$$\Gamma_3^0 = \{M \in Sp(3, \mathbf{Z}) | \; M = [S, U], \; S = {}^tS \in M_3(\mathbf{Z}),$$
$$U, U^{-1} \in GL_3(Z)\} .$$

Note that Γ_3^2, Γ_3^1 and Γ_3^0 are the stabilizers in Sp(3, Z) of the three-dimensional cusp, one-dimensional cusp and zero-dimensional cusp (or boundary component as in [24]) which we shall denote by

$$F_2 = \left\{ \begin{bmatrix} z_1 & z_{12} & * \\ z_{12} & z_2 & * \\ * & * & i^\infty \end{bmatrix} \middle| \operatorname{Im} \begin{vmatrix} z_1 & z_{12} \\ z_{12} & z_2 \end{vmatrix} > 0 \right\},$$

$$F_1 = \left\{ \begin{bmatrix} z_1 & * & * \\ * & i\infty & * \\ * & * & i\infty \end{bmatrix} \middle| \operatorname{Im} z_1 > 0 \right\}$$

and

$$F_0 = \{[i\infty]\}$$

respectively.

It is well-known that any cusp of Sp(3, **Z**) is equivalent by an element of Sp(3, **Z**) to F_2 or F_0 (p.28 of [13]). On the other hand, each element in $Sp(3, Z)$, with property (2) or (3) or (4) or (5) in Theorem 1, does stabilize at least one cusp of $Sp(3, Z)$. Thus we have the following Lemma:

LEMMA 1. *Suppose* $M \in Sp(3, Z)$ *with the property (2) or (3) or (4) or (5) as in Theorem 1, then* M *is conjugate in* $sp(3, Z)$ *to an element of* $\Gamma_3^2 \cup \Gamma_3^1 \cup \Gamma_3^0$.

With the above Lemma, the problem of finding conjugacy classes of $Sp(3, \mathbf{Z})$ now is easy since we know all the conjugacy classes of $Sp(2, \mathbf{Z})$. In this way, we get a complete list of conjugacy classes of $Sp(3, Z)$. The conjugacy classes of modular groups of higher degrees can be obtained in the same way.

2.3 Conjugacy classes of Γ_3^2

Each element in Γ_3^2 can be decomposed uniquely into a product of the form

$$\begin{bmatrix} P & 0 & Q & 0 \\ 0 & 1 & 0 & 0 \\ R & 0 & S & 0 \\ 0 & 0 & 0 & 1 \end{bmatrix} \begin{bmatrix} E & 0 & 0 & q \\ {}^t p & 1 & {}^t q & {}^t pq \\ 0 & 0 & E & -p \\ 0 & 0 & 0 & 1 \end{bmatrix} \begin{bmatrix} E & 0 & 0 & 0 \\ 0 & 1 & 0 & s \\ 0 & 0 & E & 0 \\ 0 & 0 & 0 & 1 \end{bmatrix},$$

where p and q are 1×2 row vectors, s is an integer and $M = \begin{bmatrix} P & Q \\ R & S \end{bmatrix} \in Sp(2, \mathbf{Z})$. To save space, we denote such an element by $[M; p, q; s]$.

LEMMA 2. *Suppose* $M \in Sp(3, \mathbf{Z})$ *is conjugate in*

$Sp(3, \mathbb{R})$ <u>to</u> $[\lambda_1, \lambda_2] \times \begin{bmatrix} 1 & s \\ 0 & 1 \end{bmatrix}$, <u>with</u> λ_1, λ_2 <u>roots of unity</u>, $\lambda_1^2, \lambda_1\lambda_2, \lambda_2^2 \neq 1$, <u>then</u> M <u>is conjugate in</u> $Sp(3, \mathbf{Z})$ <u>to an element of</u> Γ_3^2 .

<u>Proof</u>. The characteristic polynomial of M is a product of $[X^2 - (\lambda_1 + \bar{\lambda}_1)X + 1][X^2 - (\lambda_2 + \bar{\lambda}_2)X + 1]$ and $(X-1)^2$, hence there exists $R \in Sp(3, \mathbf{Q})$ such that

$$RMR^{-1} = M_1 \times M_2 \in Sp(2, \mathbf{Q}) \times SL_2(Q)$$

by Lemma 1 of [6]. The corresponding M_2 has characteristic polynomial $(X-1)^2$, hence M_2 is conjugate in $SL_2(Q)$ to a unipotent element $\begin{bmatrix} 1 & s \\ 0 & 1 \end{bmatrix}$. Thus M is conjugate in $Sp(3, \mathbf{Q})$ to $M_1 \times \begin{vmatrix} 1 & s \\ 0 & 1 \end{vmatrix}$.

Applying Satz 1 of [5], we conclude that M is conjugate in $Sp(3, \mathbf{Z})$ to an element of Γ_3^2 .

<u>LEMMA 3</u>. <u>Let</u> $L_1 = [M; p, q; s]$ <u>and</u> $L = [E; m, n; 0]$ <u>be elements of</u> Γ_3^2 <u>with</u> $M = \begin{bmatrix} P & Q \\ R & S \end{bmatrix}$. <u>Then</u> $LL_1L^{-1} = [M; a, b; s']$ <u>where</u>

$$a = {}^tPm + {}^tRn - m + p ,$$

$$b = {}^tQm + {}^tSn - n + q ,$$

$$s' = s + ({}^tmR + {}^tnS)(q-n) + ({}^tmQ + {}^tnS)(m-p) - {}^tpn + {}^tqm + {}^tpq .$$

<u>Proof</u>. It follows form direct calculation.

COROLLARY. *Let* $L = [M; p, q; s]$ *be an element of* Γ_3^2. *Suppose* M *has an isolated fixed point, and let* $\phi(x)$ *be its characteristic polynomial; then* L *is conjugate in* $Sp(3, Z)$ *to* $M \times \begin{bmatrix} 1 & s \\ 0 & 1 \end{bmatrix}$, $s \in \mathbf{Z}$ *if* $\phi(1) = \pm 1$.

Proof. With notations as in Lemma 3, we have

$$\phi(1) = \det \begin{bmatrix} {}^tP-E & {}^tR \\ {}^tQ & {}^tS-E \end{bmatrix} .$$

If $\phi(1) = \pm 1$, then there exists integral solutions m, n, such that $a = 0$ and $b = 0$. Thus follows our assertion.

There are 22 regular elliptic conjugacy classes in $Sp(2, \mathbf{Z})$ as shown in [11, 16]. Now fix a conjugacy class M with characteristic polynomial $\phi(X)$ and let $L = [M; p, q; s]$ be an element of Γ_2^3. With a conjugation by an element of the forms $[\pm E; m, n; 0]$, we can reduce the entries of p and q to nonegative integers which are no larger than $\frac{1}{2}|\phi(1)|^{\frac{1}{2}}$ or $\frac{1}{2}|\phi(1)|$, depending on whether $\phi(1)$ is a complete square or not. From these reduced sets, we can write down all representatives of conjugacy classes.

TABLE III Conjugacy Classes of Γ_3^2 *

Here $a \sim b$ means a is conjugate in Sp(3, Q) to b

* We exclude conjugacy classes in $\Gamma_3^2 \cap (\Gamma_3^2 \cup \Gamma_0^1)$.

---

1. $$M_1 = \begin{bmatrix} 0 & 0 & 1 & 0 \\ 0 & 0 & 0 & 1 \\ -1 & 0 & 0 & 0 \\ 0 & -1 & 0 & 0 \end{bmatrix} \sim e[1/2,\ 1/2,\ 0],\ \phi(X) = (X^2+1)^2 .$$

1-1 $[M_1;\ 0,\ 0;\ 0] \sim e[1/2,\ 1/2,\ 0]$.

1-2 $[M_1;\ 0,\ 0;\ s]$, $s \in \mathbf{Z} - \{0\}$.

1-3 $[-M_1;\ 0,\ 0;\ 0] \sim e[3/2,\ 3/2,\ 0]$.

1-4 $[-M_1;\ 0,\ 0;\ s]$, $s \in \mathbf{Z} - \{0\}$.

1-5 $[M_1;\ {}^t[1,0],\ 0;\ s] \sim [M_1;\ 0,\ 0;\ s - \frac{1}{2}]$, $s \in \mathbf{Z}$.

1-6 $[-M_1;\ {}^t[1,0],\ 0;\ s] \sim [-M_1;\ 0,\ 0;\ s + \frac{1}{2}]$, $s \in \mathbf{Z}$.

1-7 $[M_1;\ {}^t[1,1],\ 0;\ 1] \sim [M_1;\ 0,\ 0;\ 0]$.

1-8 $[M_1;\ {}^t[1,1],\ 0;\ s] \sim [M_1;\ 0,\ 0;\ s-1]$, $s \in \mathbf{Z} - \{1\}$.

1-9 $[-M_1;\ {}^t[1,1],\ 0;\ s] \sim [-M_1;\ 0,\ 0;\ s+1]$, $s \in \mathbf{Z} - \{-1\}$.

1-10 $[-M_1;\ {}^t[1,1],\ 0;\ s] \sim [-M_1;\ 0,\ 0;\ s+1]$, $s \in \mathbf{Z} - \{-1\}$.

2. $$M_2 = \begin{bmatrix} 1 & 0 & -1 & 0 \\ 0 & 1 & 0 & -1 \\ 1 & 0 & 0 & 0 \\ 0 & 1 & 0 & 0 \end{bmatrix} \sim e[5/3,\ 5/3],\ \phi(X) = (X^2-X+1)^2 .$$

2-1 $[M_2;\ 0,\ 0;\ 0] \sim e[5/3,\ 5/3,\ 0]$,

2-2 $[M_2^5;\ 0,\ 0;\ 0] \sim e[1/3,\ 1/3,\ 0]$,

2-3 $[M_2;\ 0,\ 0;\ s]$, $s \in \mathbf{Z} - \{0\}$,

2-4 $[M_2^5;\ 0,\ 0;\ s]$, $s \in \mathbf{Z} - \{0\}$,

3. $$M_3 = \begin{bmatrix} 0 & 0 & 1 & 0 \\ 0 & 0 & 0 & 1 \\ -1 & 0 & -1 & 0 \\ 0 & -1 & 0 & -1 \end{bmatrix} \sim e[2/3,\ 2/3],\ \phi(X) = (X^2+X+1)^2 .$$

3-1 $[M_3;\ 0,\ 0;\ 0] \sim e[2/3,\ 2/3,\ 0]$.

3-2 $[M_3^2;\ 0,\ 0;\ 0] \sim e[4/3,\ 4/3,\ 0]$.

3-3 $[M_3;\ 0,\ 0;\ s] \sim s \in \mathbf{Z} - \{0\}$.

3-4 $[M_3;\ {}^t[1,0],\ 0;\ s] \sim [M_3;\ 0,\ 0;\ s - \frac{1}{3}],\ s \in \mathbf{Z}$.

3-5 $[M_3;\ {}^t[1,1],\ 0;\ s] \sim [M_3;\ 0,\ 0;\ s - \frac{2}{3}],\ s \in \mathbf{Z}$.

3-6 $[M_3^2;\ 0,\ 0;\ s],\ \ s \in \mathbf{Z} - \{0\}$.

3-7 $[M_3^2;\ {}^t[1,0],\ 0;\ s] \sim [M_3^2;\ 0,\ 0;\ s + \frac{1}{3}],\ \ s \in \mathbf{Z}$.

3-8 $[M_3^2;\ {}^t[1,1],\ 0;\ s] \sim [M_3^2;\ 0,\ 0;\ s + \frac{2}{3}],\ \ s \in \mathbf{Z}$.

4. $$M_4 = \begin{bmatrix} 0 & -1 & 0 & 0 \\ 0 & 0 & -1 & 0 \\ 0 & 0 & 0 & -1 \\ 1 & 0 & 0 & 0 \end{bmatrix} \sim e[1/4,\ 5/4],\ \ \phi(X) = X^4+1 .$$

4-1 $[M_4;\ 0,\ 0;\ 0] \sim e[1/4,\ 5/4,\ 0]$.

4-2 $[M_4^3;\ 0,\ 0;\ 0] \sim e[3/4,\ 7/4,\ 0]$.

4-3 $[M_4;\ 0,\ 0;\ s],\ \ s \in \mathbf{Z} - \{0\}$.

4-4 $[M_4;\ {}^t[1,0],\ 0;\ s] \sim [M_4;\ 0,\ 0;\ s - \frac{1}{2}]$.

4-5 $[M_4;\ {}^t[0,1],\ 0;\ s] \sim [M_4;\ 0,\ 0;\ s - \frac{1}{2}]$.

4-6 $[M_4^3;\ 0,\ 0;\ s],\ \ s \in \mathbf{Z} - \{0\}$.

4-7 $[M_4^3;\ {}^t[1,0],\ 0;\ s] \sim [M_4^3;\ 0,\ 0;\ s + \frac{1}{2}],\ \ s \in \mathbf{Z}$.

4-8 $[M_4^3;\ {}^t[0,1],\ 0;\ s] \sim [M_4^3;\ 0,\ 0;\ s + \frac{1}{2}],\ \ s \in \mathbf{Z}$.

5. $$M_5 = \begin{bmatrix} 0 & -1 & -1 & 0 \\ -1 & 1 & 0 & -1 \\ 1 & -1 & -1 & 0 \\ 0 & 1 & 0 & 0 \end{bmatrix} \sim e[1/4,\ 3/4], \quad \phi(X) = X^4+1 .$$

5-1 $[M_5;\ 0,\ 0;\ 0] \sim e[1/4,\ 3/4,\ 0]$.

5-2 $[-M_5;\ 0,\ 0;\ 0] \sim e[1/4,\ 3/4,\ 1]$.

5-3 $[M_5;\ 0,\ 0;\ s]$, $s \in \mathbf{Z} - \{0\}$.

5-4 $[M_5;\ {}^t[1,0],\ 0;\ s] \sim [M_5;\ 0,\ 0;\ s+\frac{1}{2}]$, $s \in \mathbf{Z}$.

5-5 $[M_5;\ {}^t[0,1],\ 0;\ s] \sim [M_5;\ 0,\ 0;\ s+\frac{3}{2}]$, $s \in \mathbf{Z}$.

5-6 $[-M_5;\ 0,\ 0;\ s]$, $s \in \mathbf{Z} - \{0\}$.

5-7 $[-M_5;\ {}^t[1,0],\ 0;\ s] \sim [-M_5;\ 0,\ 0;\ s-\frac{3}{2}]$, $s \in \mathbf{Z}$.

5-8 $[-M_5;\ {}^t[0,1],\ 0;\ s] \sim [-M_5;\ 0,\ 0;\ s-\frac{1}{2}]$, $s \in \mathbf{Z}$.

6. $$M_6 = \begin{bmatrix} 0 & 0 & -1 & 0 \\ 0 & 0 & -1 & -1 \\ 1 & -1 & 0 & 0 \\ 0 & 1 & 0 & 0 \end{bmatrix} \sim e[1/3,\ 2/3], \quad \phi(X) = X^4+X^2+1 .$$

6-1 $[M_6;\ 0,\ 0;\ 0] \sim e[1/3,\ 2/3,\ 0]$.

6-2 $[-M_6;\ 0,\ 0;\ 0] \sim e[1/3,\ 2/3,\ 1]$.

6-3 $[M_6;\ 0,\ 0;\ s]$, $s \in \mathbf{Z} - \{0\}$.

6-4 $[M_6;\ {}^t[1,0],\ 0;\ s] \sim [M_6;\ 0,\ 0;\ s+\frac{2}{3}]$, $s \in \mathbf{Z}$.

6-5 $[-M_6;\ 0,\ 0;\ s]$, $s \in \mathbf{Z} - \{0\}$.

6-6 $[-M_6;\ {}^t[1,0],\ 0,\ s] \sim [-M_6;\ 0,\ 0;\ s-\frac{2}{3}]$, $s \in \mathbf{Z}$.

7. $$M_7 = \begin{bmatrix} 0 & 0 & -1 & 0 \\ 0 & -1 & 0 & -1 \\ 1 & 0 & 1 & 0 \\ 0 & 1 & 0 & 0 \end{bmatrix} \sim e[4/3,\ 5/3], \quad \phi(X) = X^4+X^2+1 .$$

7-1 $[M_7;\ 0,\ 0;\ 0] \sim e[4/3,\ 5/3,\ 0]$.

7-2 $[-M_7;\ 0,\ 0;\ 0] \sim e[1/3,\ 2/3,\ 0]$.

7-3 $[M_7;\ 0,\ 0;\ s]$, $s \in \mathbf{Z} - \{0\}$.

7-4 $[M_7;\ {}^t[0,1],\ 0;\ s] \sim [M_7;\ 0,\ 0;\ s+\frac{1}{3}]$.

7-5 $[-M_7;\ 0,\ 0;\ s]$, $s \in \mathbf{Z} - \{0\}$.

7-6 $[-M_7;\ {}^t[0,1],\ 0;\ s] \sim [-M_7;\ 0,\ 0;\ s-\frac{1}{3}]$.

8. $$M_8 = \begin{bmatrix} 0 & -1 & 0 & -1 \\ 1 & 0 & 0 & 0 \\ 0 & 1 & 0 & 0 \\ 0 & 0 & 1 & 0 \end{bmatrix} \sim e[2/3,\ 5/3], \quad \phi(X) = X^4+X^2+1$$

8-1 $[M_8;\ 0,\ 0;\ 0] \sim e[2/3,\ 5/3,\ 0]$.

8-2 $[M_8;\ 0,\ 0;\ s]$, $s \in \mathbf{Z} - \{0\}$.

8-3 $[M_8;\ {}^t[0,1],\ 0;\ s] \sim [M_8;\ 0,\ 0;\ s+\frac{1}{3}]$.

9. $$M_9 = \begin{bmatrix} 0 & 1 & 0 & 0 \\ 0 & 0 & 1 & 0 \\ 0 & 0 & 0 & 1 \\ -1 & 0 & -1 & 0 \end{bmatrix} \sim e[4/3,\ 1/3], \quad \phi(X) = X^4+X^2+1 .$$

9-1 $[M_9;\ 0,\ 0;\ 0] \sim e[4/3,\ 1/3,\ 0]$.

9-2 $[M_9;\ 0,\ 0;\ s]$, $s \in \mathbf{Z} - \{0\}$.

9-3 $[M_9;\ {}^t[1,0],\ 0;\ s] \sim [M_9;\ 0,\ 0;\ s+\frac{1}{3}]$, $s \in \mathbf{Z}$.

10. $M_{10} = \begin{bmatrix} -1 & 0 & -1 & 0 \\ 0 & 1 & 0 & 1 \\ 1 & 0 & 0 & 0 \\ 0 & -1 & 0 & 0 \end{bmatrix} \sim e[1/3,\ 4/3],\ \phi(X) = X^4+X^2+1$.

10-1 $[M_{10};\ 0,\ 0;\ 0] \sim e[1/3,\ 4/3,\ 0]$.

10-2 $[M_{10};\ 0,\ 0;\ s],\quad s \in \mathbf{Z} - \{0\}$.

10-3 $[M_{10};\ {}^t[1,0],\ 0;\ s] \sim [M_{10};\ 0,\ 0;\ s+\frac{1}{3}],\quad s \in \mathbf{Z}$.

11. $M_{11} = \begin{bmatrix} 0 & 0 & 1 & 0 \\ 0 & 0 & 0 & -1 \\ -1 & 0 & -1 & 0 \\ 0 & 1 & 0 & 1 \end{bmatrix} \sim e[2/3,\ 5/3],\ \phi(X) = X^4+X^2+1$.

11-1 $[M_{11};\ 0,\ 0;\ 0] \sim e[2/3,\ 5/3,\ 0]$.

11-2 $[M_{11};\ 0,\ 0;\ s],\quad s \in \mathbf{Z} - \{0\}$.

11-3 $[M_{11};\ {}^t[1,0],\ 0;\ s] \sim [M_{11};\ 0,\ 0;\ s-\frac{1}{3}]$.

12. $M_{12} = \begin{bmatrix} 0 & 0 & 0 & -1 \\ 1 & 0 & 0 & 0 \\ 0 & 1 & 0 & 1 \\ 0 & 0 & 1 & 0 \end{bmatrix} \sim e[5/6,\ 11/6],\ \phi(X) = X^4-X^2+1$.

12-1 $[M_{12};\ 0,\ 0;\ 0] \sim e[5/6,\ 11/6,\ 0]$.

12-2 $[M_{12}^5;\ 0,\ 0;\ 0] \sim e[1/6,\ 7/6,\ 0]$.

12-3 $[M_{12};\ 0,\ 0;\ s],\quad s \in \mathbf{Z} - \{0\}$.

12-4 $[M_{12}^5;\ 0,\ 0;\ s],\quad s \in \mathbf{Z} - \{0\}$.

13. $M_{13} = \begin{bmatrix} 0 & 0 & 1 & 0 \\ 0 & 0 & 0 & 1 \\ -1 & 0 & 0 & 0 \\ 0 & -1 & 0 & -1 \end{bmatrix} \sim e[1/2,\ 2/3],\ \phi(X) = (X^2+1)(X^2+X+1)$

13-1 $[M_{13};\ 0,\ 0;\ 0] \sim e[1/2,\ 2/3,\ 0]$.

13-2 $[M_{13};\ 0,\ 0;\ s],\quad s \in \mathbf{Z} - \{0\}$.

13-3 $[M_{13};\ {}^t[0,1],\ 0;\ s] \sim [M_{13};\ 0,\ 0.\ s - \frac{1}{3}],\ s \in \mathbf{Z}$.

13-4 $[M_{13};\ {}^t[1,0],\ 0;\ s] \sim [M_{13};\ 0,\ 0;\ s - \frac{1}{2}],\ s \in \mathbf{Z}$.

13-5 $[M_{13};\ {}^t[1,1],\ 0;\ s] \sim [M_{13};\ 0,\ 0;\ s - \frac{5}{6}],\ s \in \mathbf{Z}$.

13-6 $[-M_{13};\ 0,\ 0;\ 0] \sim e[1/2,\ 2/3,\ 1]$.

13-7 $[-M_{13};\ 0,\ 0;\ s],\quad s \in \mathbf{Z} - \{0\}$.

13-8 $[-M_{13};\ {}^t[1,0],\ 0;\ s] \sim [-M_{13};\ 0,\ 0;\ s + \frac{1}{2}]$.

14. $M_{14} = \begin{bmatrix} 0 & 0 & 1 & 0 \\ 0 & -1 & 0 & -1 \\ -1 & 0 & 0 & 0 \\ 0 & 1 & 0 & 0 \end{bmatrix} \sim e[1/2,\ 4/3],\ \phi(X) = (X^2+1)(X^2+X+1)$

14-1 $[M_{14};\ 0,\ 0;\ 0] \sim e[1/2,\ 4/3,\ 0]$.

14-2 $[M_{14};\ 0,\ 0;\ s],\quad s \in \mathbf{Z}$.

14-3 $[M_{14};\ {}^t[0,1],\ 0;\ s] \sim [M_{14};\ 0,\ 0;\ s + \frac{1}{3}],\ s \in \mathbf{Z}$.

14-4 $[M_{14};\ {}^t[1,0],\ 0;\ s] \sim [M_{14};\ 0,\ 0;\ s - \frac{1}{2}],\ s \in \mathbf{Z}$.

14-5 $[M_{14};\ {}^t[1,1],\ 0;\ s] \sim [M_{14};\ 0,\ 0;\ s - \frac{1}{6}],\ s \in \mathbf{Z}$.

14-6 $[-M_{14};\ 0,\ 0;\ 0] \sim e[1/2,\ 4/3,\ 1]$.

14-7 $[-M_{14};\ 0,\ 0;\ s],\quad s \in \mathbf{Z} - \{0\}$.

14-8 $[-M_{14};\ {}^t[1,0],\ 0;\ s] \sim [-M_{14};\ 0,\ 0;\ s + \frac{1}{2}]$.

15. $$M_{15} = \begin{bmatrix} 0 & 0 & 1 & 0 \\ 0 & 1 & 0 & 1 \\ -1 & 0 & 0 & 0 \\ 0 & -1 & 0 & 0 \end{bmatrix} \sim e[1/2,\ 1/3],\ \phi(X) = (X^2+1)(X^2-X+1)$$

15-1 $[M_{15};\ 0,\ 0;\ 0] \sim e[1/2,\ 1/3,\ 0]$.

15-2 $[M_{15};\ 0,\ 0;\ s],\quad s \in \mathbf{Z} - \{0\}$.

15-3 $[M_{15};\ {}^t[1,0],\ 0;\ s] \sim [M_{15};\ 0,\ 0;\ s-\frac{1}{2}],\ s \in \mathbf{Z}$.

15-4 $[-M_{15};\ 0,\ 0;\ 0] \sim e[1/2,\ 1/3,\ 1]$.

15-5 $[-M_{15};\ 0,\ 0;\ s],\ s \in \mathbf{Z} - \{0\}$.

15-6 $[-M_{15};\ {}^t[1,0],\ 0;\ s] \sim [M_{15};\ 0,\ 0;\ s+\frac{1}{2}]$.

15-7 $[-M_{15};\ {}^t[0,1],\ 0;\ s] \sim [-M_{15};\ 0,\ 0;\ s+\frac{2}{3}]$.

15-8 $[-M_{15};\ {}^t[1,1],\ 0;\ s] \sim [-M_{15};\ 0,\ 0;\ s+\frac{7}{6}]$.

16. $$M_{16} = \begin{bmatrix} 0 & 0 & 1 & 0 \\ 0 & 1 & 0 & -1 \\ -1 & 0 & 0 & 0 \\ 0 & 1 & 0 & 1 \end{bmatrix} \sim e[1/2,\ 5/3],\ \phi(X) = (X^2+1)(X^2-X+1)$$

16-1 $[M_{16};\ 0,\ 0;\ 0] \sim e[1/2,\ 5/3,\ 0]$.

16-2 $[M_{16};\ 0,\ 0;\ s],\quad s \in \mathbf{Z} - \{0\}$.

16-3 $[M_{16};\ {}^t[1,0],\ 0;\ s] \sim [M_{16};\ 0,\ 01\ s-\frac{1}{2}],\quad s \in \mathbf{Z}$.

16-4 $[-M_{16};\ 0,\ 0;\ 0] \sim e[1/2,\ 5/3,\ 1]$.

16-5 $[-M_{15};\ 0,\ 0;\ s],\quad s \in \mathbf{Z} - \{0\}$.

16-6 $[-M_{16};\ {}^t[1,0],\ 0;\ s] \sim [-M_{16};\ 0,\ 0;\ s+\frac{1}{2}]$.

16-7 $[-M_{16};\ {}^t[0,1],\ 0;\ s] \sim [-M_{16};\ 0,\ 0;\ s+\frac{1}{3}]$.

16-8 $[-M_{16};\ {}^t[1,1],\ 0;\ s] \sim [-M_{16};\ 0,\ 0;\ s+\frac{5}{6}]$.

17. $$M_{17} = \begin{bmatrix} -1 & 0 & 1 & 0 \\ 0 & 0 & 0 & 1 \\ 1 & 0 & 0 & -1 \\ 1 & -1 & -1 & 0 \end{bmatrix} \sim e[2/5,\ 4/5],\ \phi(X) = X^4+X^3+X^2+X+1$$

17-1 $[M_{17};\ 0,\ 0;\ 0] \sim e[2/5,\ 4/5,\ 0]$.

17-2 $[-M_{17};\ 0,\ 0;\ 0] \sim e[2/5,\ 4/5,\ 1]$.

17-3 $[M_{17}^2;\ 0,\ 0;\ 0] \sim e[4/5,\ 8/5,\ 0]$.

17-4 $[-M_{17}^2;\ 0,\ 0;\ 0] \sim e[4/5,\ 8/5,\ 1]$.

17-5 $[M_{17}^3;\ 0,\ 0;\ 0] \sim e[6/5,\ 2/5,\ 0]$.

17-6 $[-M_{17}^3;\ 0,\ 0;\ 0] \sim e[6/5,\ 2/5,\ 1]$.

17-7 $[M_{17}^4;\ 0,\ 0;\ 0] \sim e[8/5,\ 6/5,\ 0]$.

17-8 $[-M_{17}^4;\ 0,\ 0;\ 0] \sim e[8/5,\ 6/5,\ 1]$.

17-9 $[M_{17};\ 0,\ 0;\ s]$, $s \in \mathbf{Z} - \{0\}$.

17-10 $[M_{17};\ {}^t[1,0],\ 0;\ s] \sim [M_{17};\ 0,\ 0;\ s - \frac{2}{5}]$, $s \in \mathbf{Z}$.

17-11 $[M_{17};\ {}^t[2,0],\ 0;\ s] \sim [M_{17};\ 0,\ 0;\ s - \frac{8}{5}]$, $s \in \mathbf{Z}$.

17-12 $[-M_{17};\ 0,\ 0;\ s]$, $s \in \mathbf{Z} - \{0\}$.

17-13 $[M_{17}^2;\ 0,\ 0;\ s]$, $s \in \mathbf{Z} - \{0\}$.

17-14 $[M_{17}^2;\ {}^t[1,0],\ 0;\ s] \sim [M_{17}^2;\ 0,\ 0;\ s + \frac{2}{5}]$, $s \in \mathbf{Z}$.

17-15 $[M_{17}^2;\ {}^t[2,0],\ 0;\ s] \sim [M_{17}^2;\ 0,\ 0;\ s + \frac{8}{5}]$, $s \in \mathbf{Z}$.

17-16 $[-M_{17}^2;\ 0,\ 0;\ s]$, $s \in \mathbf{Z} - \{0\}$.

17-17 $[M_{17}^3;\ 0,\ 0;\ s]$, $s \in \mathbf{Z} - \{0\}$.

17-18 $[M_{17}^3;\ {}^t[1,0],\ 0;\ s] \sim [M_{17}^3;\ 0,\ 0;\ s + \frac{3}{5}]$, $s \in \mathbf{Z}$.

17-19 $[M_{17}^3;\ {}^t[2,0],\ 0;\ s] \sim [M_{17}^3;\ 0,\ 0;\ s + \frac{12}{5}]$, $s \in \mathbf{Z}$.

17-20 $[-M_{17}^3;\ 0,\ 0;\ s]$, $s \in \mathbf{Z} - \{0\}$.

17-21 $[M^4_{17};\ 0,\ 0;\ s]$, $s \in \mathbf{Z} - \{0\}$.

17-22 $[M^4_{17};\ {}^t[1,0],\ 0;\ s] \sim [M^4_{17};\ 0,\ 0;\ s+\tfrac{1}{2}]$, $s \in \mathbf{Z}$.

17-23 $[M^4_{17};\ {}^t[2,0],\ 0;\ s] \sim [M^4_{17};\ 0,\ 0;\ s+\tfrac{1}{2}]$, $s \in \mathbf{Z}$.

17-24 $[-M^4_{17}\ 0,\ 0;\ s]$, $s \in \mathbf{Z} - \{0\}$.

2.4 Conjugacy classes of Γ^1_3 .

Each element L in Γ^1_3 can be factored uniquely into a product of the form

$$L = \begin{bmatrix} a & 0 & b & 0 \\ 0 & U & 0 & 0 \\ c & 0 & d & 0 \\ 0 & 0 & 0 & {}^tU^{-1} \end{bmatrix} \begin{bmatrix} 1 & 0 & 0 & {}^tq \\ p & E_2 & q & p\,{}^tq \\ 0 & 0 & 1 & -{}^tp \\ 0 & 0 & 0 & E_2 \end{bmatrix} \begin{bmatrix} 1 & 0 & 0 & 0 \\ 0 & E_2 & 0 & S \\ 0 & 0 & 1 & 0 \\ 0 & 0 & 0 & E_2 \end{bmatrix} .$$

Here p,q are 1×2 row vectors and S is a 2×2 symmetric matrix. To save spaces, we denote such L by

$$[M,\ U;\ p,\ q;\ S]$$

with $M = \begin{bmatrix} a & b \\ c & d \end{bmatrix}$.

LEMMA 4. *Let* $L = [M,\ U;\ p,\ q;\ S]$ *and* $L_1 = [E_2,\ E_2;\ m,\ n;\ 0]$ *be elements of* Γ^1_3 . *If* $M = \begin{bmatrix} a & b \\ c & d \end{bmatrix}$, *then*

$$L_1 L L_1^{-1} = [M,\ U;\ u,\ v;\ T]$$

with

$$u = p + U^{-1}(am + cn) - m ,$$

$$v = q + U^{-1}(bm + dm) - n ,$$

$$T = U^{-1}(am + cn)({}^tq - {}^tn) + U^{-1}(bm + dn)({}^tm - {}^tp)$$

$$+ S + p{}^tq - p{}^tn + q{}^tm - n{}^tm + U^{-1}(m{}^tn){}^tU^{-1} .$$

COROLLARY. Let $L = [M, U; p, q; s]$ be an element of Γ_3^1 with $M = \begin{bmatrix} a & b \\ c & d \end{bmatrix}$. If

$$\det \begin{bmatrix} aU^{-1}-E_2 & cU^{-1} \\ bU^{-1} & dU^{-1}-E_2 \end{bmatrix} \neq 0 ,$$

then L is conjugate in $Sp(3, Q)$ to an $M \times [S;, U]$ in $SL_2(Q) \times Sp(2, Q)$.

REMARK: The determinant in the above corollary is zero only when $M = U$.

TABLE IV conjugacy classes of Γ_3^1

Here $M_1 = \begin{bmatrix} 0 & 1 \\ -1 & 0 \end{bmatrix}$, $M_2 = \begin{bmatrix} 1 & -1 \\ 1 & 0 \end{bmatrix}$, $M_3 = \begin{bmatrix} 0 & -1 \\ 1 & -1 \end{bmatrix}$, $U_1 = \begin{bmatrix} 1 & 0 \\ 0 & -1 \end{bmatrix}$ and $U_2 = \begin{bmatrix} 1 & 1 \\ 0 & -1 \end{bmatrix}$. $a \sim b$ means a is conjugate in $Sp(3, Q)$ to b.

1-1 $[\pm M_1, E_2; 0, 0; 0] = (\pm M_1) \times E_4$ (two conjugacy classes).

1-2 $[\pm M_1, E_2; 0, 0; \begin{bmatrix} s & 0 \\ 0 & 0 \end{bmatrix}]$, $s \in \mathbf{Z} - \{0\}$.

1-3 $[\pm M_1, E_2; 0, 0; S]$, $S = {}^tS \in M_2(\mathbf{Z})$, $\det S \neq 0$.

1-4 $[\pm M_1, E_2; \begin{bmatrix} 1 \\ 0 \end{bmatrix}, 0; \begin{bmatrix} s & 0 \\ 0 & 0 \end{bmatrix}] \sim$

$[\pm M_1, E_2; 0, 0; \begin{bmatrix} s\mp\frac{1}{2} & 0 \\ 0 & 0 \end{bmatrix}]$, $s \in \mathbf{Z}$.

1-5 $[\pm M_1, E_2; \begin{bmatrix} 1 \\ 0 \end{bmatrix}, 0; S] \sim [\pm M_1, E_2; 0, 0; \begin{bmatrix} s_1\mp\frac{1}{2} & s_{12} \\ s_{12} & s_2 \end{bmatrix}]$,

$\det S \neq 0$ or $s_1 = s_{12} = 0$, $s_2 \neq 0$.

1-6 $[\pm M_1, M_2; 0, 0; 0]$.

1-7 $[\pm M_1, M_2; 0, 0; \begin{bmatrix} s & 0 \\ 0 & 0 \end{bmatrix}]$, $s \in \mathbf{Z} - \{0\}$.

1-8 $[M_1\ U_1; 0, 0; 0]$.

1-9 $[M_1, U_1; 0, 0; \begin{bmatrix} 0 & 1 \\ 1 & 0 \end{bmatrix}] \sim [M_1, U_1; 0, 0; 0]$.

1-10 $[\pm M_1, U_1; 0, 0; \begin{bmatrix} s & 0 \\ 0 & 0 \end{bmatrix}]$, $s \in \mathbf{Z} - \{0\}$.

1-11 $[\pm M_1,\ U_1;\ 0,\ 0;\ \begin{bmatrix} s & 1 \\ 1 & 0 \end{bmatrix}],\quad s \in Z - \{0\}$.

1-12 $[M_1,\ U_1;\ 0,\ 0;\ \begin{bmatrix} s_1 & 0 \\ 0 & s_2 \end{bmatrix}],\quad s_1, s_2 \in Z - \{0\}$.

1-13 $[M_1,\ U_1;\ 0,\ 0;\ \begin{bmatrix} s_1 & 1 \\ 1 & s_2 \end{bmatrix}],\quad s_1, s_2 \in \mathbf{Z} - \{0\}$.

1-14 $[\pm M_1,\ U_1;\ \begin{bmatrix} 1 \\ 0 \end{bmatrix},\ 0;\ \begin{bmatrix} s & 0 \\ 0 & 0 \end{bmatrix}] \sim$

$$[M_1,\ U_1;\ 0,\ 0;\ \begin{bmatrix} s-\frac{1}{2} & 0 \\ 0 & 0 \end{bmatrix}],\quad s \in \mathbf{Z} .$$

1-15 $[\pm M_1;,\ U_1;\ \begin{bmatrix} 1 \\ 0 \end{bmatrix},\ 0;\ \begin{bmatrix} s & 1 \\ 1 & 0 \end{bmatrix}],\quad s \in \mathbf{Z}$.

1-16 $[M_1,\ U_1;\ \begin{bmatrix} 1 \\ 0 \end{bmatrix},\ 0;\ \begin{bmatrix} s_1 & 0 \\ 0 & s_2 \end{bmatrix}] \sim$

$$[M_1,\ U_1;\ 0,\ 0;\ \begin{bmatrix} s_1+\frac{1}{2} & 0 \\ 0 & s_2 \end{bmatrix}],\ s_1 \in \mathbf{Z},\quad s_2 \in \mathbf{Z} - \{0\}.$$

1-17 $[M_1,\ U_1;\ \begin{bmatrix} 1 \\ 0 \end{bmatrix},\ 0;\ \begin{bmatrix} s_1 & 1 \\ 1 & s_2 \end{bmatrix}] \sim$

$$[M_1,\ U_1;\ 0,\ 0;\ \begin{bmatrix} s_1+\frac{1}{2} & 0 \\ 0 & s_2 \end{bmatrix}],\ s_1 \in Z,\quad s_2 \in \mathbf{Z} - \{0\}.$$

1-18 $[M_1,\ U_1;\ \begin{bmatrix} 1 \\ 1 \end{bmatrix},\ 0;\ \begin{bmatrix} s_1 & 0 \\ 0 & s_2 \end{bmatrix}] \sim$

$$[M_1,\ U_1;\ 0,\ 0;\ \begin{bmatrix} s_1-\frac{1}{2} & 0 \\ 0 & s+\frac{1}{2} \end{bmatrix}],\ s_1, s_2 \in \mathbf{Z} .$$

1-19 $[M_1,\ U_1;\ \begin{bmatrix} 1 \\ 1 \end{bmatrix},\ 0;\ \begin{bmatrix} s_1 & 1 \\ 1 & s_2 \end{bmatrix}] \sim$

$$[M_1,\ U_1;\ 0,\ 0;\ \begin{bmatrix} s_1-\frac{1}{2} & 0 \\ 0 & s_2+\frac{1}{2} \end{bmatrix}],\ s_1, s_2 \in Z .$$

1-20 $[M_1,\ U_2;\ 0,\ 0;\begin{bmatrix} s_1 & 0 \\ 0 & s_2 \end{bmatrix}],\quad s_1, s_2 \in Z - \{0\}$.

1-21 $[M_1, U_2; \begin{bmatrix}1\\0\end{bmatrix}, 0; \begin{bmatrix}s_1 & 0\\0 & s_2\end{bmatrix}] \sim$

$[M_1, U_2; 0, 0; \begin{bmatrix}s_1-\frac{1}{2} & 0\\0 & s_2\end{bmatrix}]$, $s_1 \in \mathbf{Z}$, $s_2 \in \mathbf{Z} - \{0\}$.

1-22 $[M_1, U_2; \begin{bmatrix}0\\1\end{bmatrix}, 0; \begin{bmatrix}s_1 & 0\\0 & s_2\end{bmatrix}] \sim$

$[M_1, U_2; 0, 0; \begin{bmatrix}s_1 & -\frac{1}{2}\\-\frac{1}{2} & s_2+\frac{1}{2}\end{bmatrix}]$, $s_1 \in \mathbf{Z} - \{0\}$, $s_2 \in \mathbf{Z}$.

1-23 $[M_1, M_1; 0, 0; 0]$.

1-24 $[M_1, M_1; 0, 0; \begin{bmatrix}0 & 1\\1 & 0\end{bmatrix}] \sim [M_1, M_1; 0, 0; 0]$.

1-25 $[M_1, M_1; 0, 0; \begin{bmatrix}s & 0\\0 & 0\end{bmatrix}]$, $s \in \mathbf{Z} - \{0\}$.

1-26 $[M_1, M_1; 0, 0; \begin{bmatrix}s & 1\\1 & 0\end{bmatrix}]$, $s \in \mathbf{Z} - \{0\}$.

1-27 $[M_1, M_1; 0, 0; 0] \cdot [S, E]$, $S = \begin{bmatrix}0 & s & 0\\s & 0 & u\\0 & u & 0\end{bmatrix}$],

$s \in \mathbf{Z} - \{0\}$, $u = 0, 1$.

1-28 $[M_1, M_1; 0, 0; 0] \cdot [S, E]$, $S = \begin{bmatrix}0 & s & 0\\s & s' & u\\0 & u & 0\end{bmatrix}$

$s, s' \in \mathbf{Z} - \{0\}$, $u = 0, 1$.

From 2-1 to 2-36, we have $j = 1$ or 5 .

2-1 $[M_2^j, E_2; 0, 0; 0]$.

2-2 $[M_2^j, E_2; 0, 0; \begin{bmatrix}s & 0\\0 & 0\end{bmatrix}]$, $s \in Z - \{0\}$.

2-3 $[M_2^j,\ E_2;\ 0,\ 0;\ S]$, $S = {}^tS \in M_2(Z)$, $\det S \neq 0$.

2-4 $[M_2^j,\ M_1;\ 0,\ 0;\ 0]$.

2-5 $[M_2^j,\ M_1;\ 0,\ 0;\ \begin{bmatrix} s & 0 \\ 0 & 0 \end{bmatrix}]$, $s \in \mathbf{Z} - \{0\}$.

2-6 $[M_2^j,\ M_1;\ 0,\ 0;\ \begin{bmatrix} 0 & 1 \\ 1 & 0 \end{bmatrix}] \sim [M_2^j,\ M_1;\ 0,\ 0;\ 0]$.

2-7 $[M_2^j,\ M_1;\ 0,\ 0;\ \begin{bmatrix} s & 1 \\ 1 & 0 \end{bmatrix}]$, $s \in \mathbf{Z} - \{0\}$.

2-8 $[M_2^j,\ -U_1;\ 0,\ 0;\ 0]$.

2-9 $[M_2^j,\ -U_1;\ 0,\ 0;\ \begin{bmatrix} s & 0 \\ 0 & 0 \end{bmatrix}]$, $s \in \mathbf{Z} - \{0\}$.

2-10 $[M_2^j,\ -U_1;\ 0,\ 0;\ \begin{bmatrix} s_1 & 0 \\ 0 & s_2 \end{bmatrix}]$, $s \in \mathbf{Z} - \{0\}$.

2-11 $[M_2^j,\ -U_1;\ 0,\ 0;\ \begin{bmatrix} s_1 & 0 \\ 0 & s_2 \end{bmatrix}]$, $s_1, s_2 \in \mathbf{Z} - \{0\}$.

2-12 $[M_2^j,\ -U_1;\ 0,\ 0;\ \begin{bmatrix} 0 & 1 \\ 1 & 0 \end{bmatrix}] \sim [M_2^j,\ -U_2;\ 0,\ 0;\ 0]$.

2-13 $[M_2^j,\ -U_1;\ 0,\ 0;\ \begin{bmatrix} s & 1 \\ 1 & 0 \end{bmatrix}]$, $s \in \mathbf{Z} - \{0\}$.

2-14 $[M_2^j,\ -U_1;\ 0,\ 0;\ \begin{bmatrix} 0 & 1 \\ 1 & s \end{bmatrix}]$, $s \in \mathbf{Z} - \{0\}$.

2-15 $[M_2^j,\ -U_1;\ 0,\ 0;\ \begin{bmatrix} s_1 & 1 \\ 1 & s_2 \end{bmatrix}]$, $s_1, s_2 \in \mathbf{Z} - \{0\}$.

2-16 $[M_2,\ -U_1;\ \begin{bmatrix} 1 \\ 0 \end{bmatrix},\ 0;\ \begin{bmatrix} s & 0 \\ 0 & 0 \end{bmatrix}] \sim [M_2,\ -U_1;\ 0,\ 0;\ \begin{bmatrix} s-\frac{1}{3} & 0 \\ 0 & 0 \end{bmatrix}]$.

2-17 $[M_2, -U_1; \begin{bmatrix}1\\0\end{bmatrix}, 0; \begin{bmatrix}s_1 & 0\\0 & s_2\end{bmatrix}] \sim [M_2, -U_1; 0, 0; \begin{bmatrix}s-\frac{1}{3} & 0\\0 & s_2\end{bmatrix}]$,

$s_1 \in \mathbf{Z}, \quad s_2 \in Z - \{0\}$.

2-18 $[M_2, -U_1; \begin{bmatrix}1\\0\end{bmatrix}, 0; \begin{bmatrix}s & 1\\1 & 0\end{bmatrix}] \sim [M_2, -U_1; 0, 0; \begin{bmatrix}s-\frac{1}{3} & 0\\0 & 0\end{bmatrix}]$,

$s \in \mathbf{Z}$.

2-19 $[M_2, -U_1; \begin{bmatrix}1\\0\end{bmatrix}, 0; \begin{bmatrix}s_1 & 1\\1 & s_2\end{bmatrix}], \quad s_1 \in \mathbf{Z}, \quad s_2 \in \mathbf{Z} - \{0\}$.

2-20 $[M_2, -U_2; 0, 0; \begin{bmatrix}s_1 & 0\\0 & s_2\end{bmatrix}], \quad s_1, s_2 \in Z - \{0\}$.

2-21 $[M_2, -U_2; \begin{bmatrix}1\\0\end{bmatrix}, 0; \begin{bmatrix}s_1 & 0\\0 & s_2\end{bmatrix}] \sim [M_2, -U_2; 0, 0; \begin{bmatrix}s_1+\frac{1}{3} & 0\\0 & s_2\end{bmatrix}]$,

$s_1 \in \mathbf{Z}, \quad s_2 \in \mathbf{Z} - \{0\}$.

2-22 $[M_2^5, -U_1; \begin{bmatrix}1\\0\end{bmatrix}, 0; \begin{bmatrix}s_1 & 0\\0 & 0\end{bmatrix}] \sim [M_2^5, -U_1; 0, 0; \begin{bmatrix}s+\frac{1}{3} & 0\\0 & 0\end{bmatrix}]$,

$s \in \mathbf{Z}$.

2-23 $[M_2^5, -U_1; \begin{bmatrix}1\\0\end{bmatrix}, 0; \begin{bmatrix}s_1 & 0\\0 & s_2\end{bmatrix}], \quad s_1 \in Z, \quad s_2 \in \mathbf{Z} - \{0\}$.

2-24 $[M_2^5, -U_1; \begin{bmatrix}1\\0\end{bmatrix}, 0; \begin{bmatrix}s & 1\\1 & 0\end{bmatrix}] \sim [M_2^5, -U_1; 0, 0; \begin{bmatrix}s+\frac{1}{3} & 0\\0 & 0\end{bmatrix}]$,

$s \in \mathbf{Z}$.

2-25 $[M_2^5, -U_1; \begin{bmatrix}1\\0\end{bmatrix}, 0; \begin{bmatrix}s_1 & 1\\1 & s_2\end{bmatrix}], \quad s_1 \in \mathbf{Z}, \quad s_2 \in Z - \{0\}$.

2-26 $[M_2^5, -U_2; 0, 0; \begin{bmatrix}s_1 & 0\\0 & s_2\end{bmatrix}], \quad s_1, s_2 \in \mathbf{Z} - \{0\}$.

2-27 $[M_2^5, -U_2; \begin{bmatrix}1\\0\end{bmatrix}, 0; \begin{bmatrix}s_1 & 0\\0 & s_2\end{bmatrix}] \sim [M_2^5, -U_2; 0, 0; \begin{bmatrix}s_1+\frac{1}{3} & 0\\0 & s_2\end{bmatrix}]$,

$s_1 \in \mathbf{Z}, \quad s_2 \in \mathbf{Z} - \{0\}$.

2-28 $[M_2^j, M_3; 0, 0; 0]$.

2-29 $[M_2^j, M_3; 0, 0; \begin{bmatrix}s & 0\\0 & 0\end{bmatrix}], \quad s \in \mathbf{Z} - \{0\}$.

2-30 $[M_2, M_3; \begin{bmatrix}1\\0\end{bmatrix}, 0; \begin{bmatrix}s & 0\\0 & 0\end{bmatrix}] \sim [M_2, M_3; 0, 0; \begin{bmatrix}s-\frac{1}{2} & 0\\0 & 0\end{bmatrix}]$,

$s \in \mathbf{Z}$.

2-31 $[M_2^5, M_3; \begin{bmatrix}1\\0\end{bmatrix}, 0; \begin{bmatrix}s & 0\\0 & 0\end{bmatrix}] \sim [M_2, M_3; 0, 0; \begin{bmatrix}s+\frac{1}{2} & 0\\0 & 0\end{bmatrix}]$,

$s \in \mathbf{Z}$.

2-32 $[M_2^j, M_2^j; 0, 0; 0]$.

2-33 $[M_2^j, M_2^j; 0, 0; \begin{bmatrix}s & 0\\0 & 0\end{bmatrix}], \quad s \in \mathbf{Z} - \{0\}$.

2-34 $[M_2^j, M_2^j; 0, 0; 0] \cdot [S, E]$ with $S = \begin{bmatrix}0 & s & 0\\s & 0 & 0\\0 & 0 & 0\end{bmatrix}, \quad s \in \mathbf{Z} - \{0\}$.

2-35 $[M_2^j, M_2^j; 0, 0; 0] \cdot [S, E]$ with $S = \begin{bmatrix}0 & s & 0\\s & 0 & 0\\0 & 0 & 0\end{bmatrix}$,

$s, s' \in \mathbf{Z} - \{0\}$.

From 3-1 to 3-7, we have $k = 1$ or 2

3-1 $[M_3^k, E_2; 0, 0; 0]$.

3-2 $[M_3^k, E_2; 0, 0; \begin{bmatrix} s & 0 \\ 0 & 0 \end{bmatrix}]$, $s \in \mathbf{Z} - \{0\}$.

3-3 $[M_3^k, E_2; 0, 0; S]$, $S = {}^tS \in M_2(\mathbf{Z})$, $\det S \neq 0$.

3-4 $[M_3, E_2; \begin{bmatrix} 1 \\ 0 \end{bmatrix}, 0; \begin{bmatrix} s & 0 \\ 0 & 0 \end{bmatrix}] \sim [M_3, E_2; 0, 0; \begin{bmatrix} s+\frac{1}{3} & 0 \\ 0 & 0 \end{bmatrix}]$, $s \in \mathbf{Z}$.

3-5 $[M_3^2, E_2; \begin{bmatrix} 1 \\ 0 \end{bmatrix}, 0; \begin{bmatrix} s & 0 \\ 0 & 0 \end{bmatrix}] \sim [M_3^2, E_2; 0, 0; \begin{bmatrix} s-\frac{1}{3} & 0 \\ 0 & 0 \end{bmatrix}]$, $s \in \mathbf{Z}$.

3-6 $[M_3, E_2; \begin{bmatrix} 1 \\ 0 \end{bmatrix}, 0; \begin{bmatrix} s_1 & s_{12} \\ s_{12} & s_2 \end{bmatrix}] \sim [M_3, E_2; 0, 0; \begin{bmatrix} s_1+\frac{1}{3} & s_{12} \\ s_{12} & s_2 \end{bmatrix}]$,

$s_1s_2 - s_{12}^2 \neq 0$ or $s_1 = s_{12} = 0$, $s_2 \neq 0$.

3-7 $[M_3^2, E_2; \begin{bmatrix} 1 \\ 0 \end{bmatrix}, 0; \begin{bmatrix} s_1 & s_{12} \\ s_{12} & s_2 \end{bmatrix}] \sim [M_3, E_2; 0, 0; \begin{bmatrix} s_1-\frac{1}{3} & s_{12} \\ s_{12} & s_2 \end{bmatrix}]$,

$s_1s_2 - s_{12}^2 \neq 0$ or $s_1 = s_{12} = 0$, $s_2 \neq 0$.

2.5 Conjugacy classes of Γ_3^0.

The conjugacy classes of Γ_3^0 are completely determined by the following Lemma.

LEMMA 5. *Suppose* $M \in Sp(3, \mathbf{Z})$ *is conjugate in* $Sp(3, \mathbf{R})$ *to* $S', U' \in Sp(3, \mathbf{R})$ *with* $U' = E_3$ *or* $\mathrm{diag}[1, 1, -1]$ *or*

$$\begin{bmatrix} 1 & 0 & 0 \\ 0 & \cos\theta & \sin\theta \\ 0 & -\sin\theta & \cos\theta \end{bmatrix}, \quad \theta = \pm\frac{\pi}{2}, \pm\frac{\pi}{3}, \pm\frac{2\pi}{3},$$

then M *is conjugate in* $Sp(3, \mathbf{Z})$ *to one of the following in the type* $[S, U]$

(1) $[S, E_3]$, $S = {}^tS \in M_3(\mathbf{Z})$,

(2) $S = \begin{bmatrix} s_1 & s_{12} & 0 \\ s_{12} & s_2 & u \\ 0 & u & s_3 \end{bmatrix}$, $U = \mathrm{diag}[1, 1, -1]$, $u = 0$ *or* 1,

(3) $S = \begin{bmatrix} s_1 & u & 0 \\ u & s_2 & v \\ 0 & v & 0 \end{bmatrix}$, $u, v \in \{0, 1\}$, $U = \begin{bmatrix} 1 & 0 & 0 \\ 0 & 0 & 1 \\ 0 & -1 & 0 \end{bmatrix}$,

(4) $S = \mathrm{diag}[s_1, s_2, 0]$, $U = \begin{bmatrix} 1 & 0 & 0 \\ 0 & 1 & -1 \\ 0 & 1 & 0 \end{bmatrix}$.

(5) $S = \begin{bmatrix} s_1 & u & 0 \\ u & s_2 & 0 \\ 0 & 0 & 0 \end{bmatrix}$, $U = \begin{bmatrix} 1 & 0 & 0 \\ 0 & -1 & 1 \\ 0 & -1 & 0 \end{bmatrix}$, $u = 0$ *or* 1,

$$(6)\qquad S = \begin{bmatrix} s_1 & s_{12} & 0 \\ s_{12} & s_2 & 0 \\ 0 & 0 & s_3 \end{bmatrix}, \quad U = \begin{bmatrix} 1 & 0 & 0 \\ 0 & 1 & 0 \\ 0 & 1 & -1 \end{bmatrix}, \quad s_{12} \ \underline{\text{odd if}} \ s_2+4s_3 = 0.$$

Proof. Here we only prove (5). Others follow with similar arguements. The characteristic polynomial of M can be factored into $(X-1)^2(X^2+X+1)^2$. By Lemma 1 of [6] and Satz 2 of [5], there exists $L_1 \in Sp(3, Z)$ such that

$$L_1^{-1}ML_1 = \begin{bmatrix} 1 & * & s & * \\ 0 & P & * & Q \\ 0 & 0 & 1 & 0 \\ 0 & R & * & S \end{bmatrix}$$

with $M' = \begin{bmatrix} P & Q \\ R & S \end{bmatrix} \in Sp(2, \mathbf{Z})$ and M' has characteristic polynomial $(X^2+X+1)^2$. It follows that M' is conjugate in $Sp(2, \mathbf{Z})$ to $[S, U]$ with $U = \begin{bmatrix} -1 & 1 \\ -1 & 0 \end{bmatrix}$ or $\begin{bmatrix} 1 & -1 \\ 1 & 0 \end{bmatrix}^m \times \begin{bmatrix} 1 & -1 \\ 1 & 0 \end{bmatrix}^m$, $m = 2$ or 4. Our assumptions insure us that the later cannot happen. Hence M is conjugate in $Sp(3, Z)$ to $[S_1, U_1]$ with

$$U_1 = \begin{bmatrix} 1 & u & v \\ 0 & -1 & 1 \\ 0 & -1 & 0 \end{bmatrix}.$$

Let $L_2 = [0, V]$ with

$$V = \begin{bmatrix} \pm 1 & p & q \\ 0 & 1 & 0 \\ 0 & 0 & 1 \end{bmatrix}, \quad p, q \in \mathbf{Z}.$$

Then $L_2[S_1, U_1]L_2^{-1} = [S_2, VU_1V^{-1}]$ with

$$VU_1V^{-1} = \begin{bmatrix} 1 & a & b \\ 0 & -1 & 1 \\ 0 & -1 & 0 \end{bmatrix}, \quad a = \pm(u-2p-q), \quad b = \pm(v+p-q) .$$

Thus we can choose suitable integers p, q so that $a = 0, b = 0$ or 1. If $a = 0$ and $b = 0$, then with a conjugation by element of the form $[T, E] \in \Gamma_3^0$, we are done. If $a = 0$ and $b = 1$, with a conjugation with $[T, E] \in \Gamma_3^0$ if necessary, we may assume that the entries of S_2 at 1, 12, 13 positions are 0. Then we have the decomposition

$$[S_2, VU_1V^{-1}] = [S_2, E][0, V_1][0, U]$$

with

$$V_1 = \begin{bmatrix} 1 & 1 & -1 \\ 0 & 1 & 0 \\ 0 & 0 & 1 \end{bmatrix} \quad \text{and} \quad U = \begin{bmatrix} 1 & 0 & 0 \\ 0 & -1 & 1 \\ 0 & -1 & 0 \end{bmatrix}.$$

Let $L_3 = \begin{vmatrix} A & B \\ -B & A \end{vmatrix}$ with $A + Bi = \text{diag}[1, i, i]$. Then a direct calculation shows

$$L_3[S_2, E]L_3^{-1} = [S_3, E],$$

$$L_3[0, V_1]L_3^{-1} = [S_4, E],$$

$$L_3[0, U]L_3^{-1} = [0, U'], \quad U' = \begin{bmatrix} 1 & 0 & 0 \\ 0 & 0 & 1 \\ 0 & -1 & -1 \end{bmatrix}.$$

It follows that M is conjugate in $Sp(3, \mathbf{Z})$ to $[S_3+S_4, U']$ which is conjugate in Γ_3^0 to $[S, U]$ in (5). This proves our assertion.

2.6 Applications and further remarks

The Selberg trace formula reduces the problem of calculating the dimension of cusp forms of Siegel upper-half space, in case the fundamental domain is not compact but has finite volume, to the evaluation of certain integrals combining with special values of certain zeta functions. When the degree is two or three, all types of these integrals are evaluated in [10, 11] or [16], and the corresponding zeta functions can be evaluated [10, 11, 16, 31] . Incoporated with the conjugacy classes of Sp(3, Z) in this CHAPTER it is expected an explicit formula for the dimension of Siegel cusp forms of degree three with respect to Sp(3, Z) can be obtained.

CHAPTER III

EXPLICIT EVALUATIONS

3.1 Introduction

Though we know all nontrivial contributions to the dimension formula are those from conjugacy classes of regular elliptic elements, Γ_3^1, Γ_3^2 and we can compute their contributions individually by results obtained in [11]; it is still far away from our final purpose — to get an explicit formula for $\dim_{\mathbf{C}} S(k; Sp(3, \mathbf{Z}))$ when k is even and sufficiently large. One main reason for this obstacle is the number of conjugacy classes in $Sp(3, \mathbf{Z})$ is too big to handle. Any mistake in the course of computation may lead to an incorrect formula.

By any means, we shall compute all nontrivial contributions and add them together to get a final formula. In our process of calculation, we shall combine certain contributions from certain selected conjugacy classes so that their total contribution is a sum of products of rational polynomials with periodic rational-valued functions in weight k. For example, the total contribution from elements of order 7 is

$$7^{-1}[1, 0, 1, 0, 0, 0, 0]$$

while the total contribution from elements with characteristic polynomial $(X-1)^2(X^4+X^3+X^2+X+1)$ or $(X-1)^2(X^4-X^3+X^2-X+1)$ is

$$2^{-3}3^{-1}5^{-2}(2k-4)[1,0,-1, 3, -3] + 2^{-3}3^{-1}5^{-2}[-66,0,54,-54,66] .$$

Here we use the notation

$$\alpha(k) = [a_0, a_1, \dots, a_{m-1}]$$

to stand for the periodic function in k of period $2m$ defined by

$$\alpha(k) = a_j \quad \text{if} \quad k \equiv 2j \pmod{2m} .$$

Once and for all, we assume the weight k is an even integer greater than 9 .

3.2 Contributions from Conjugacy Classes of Regular Elliptic Elements.

Contributions from conjugacy classes of regular elliptic elements in $Sp(n, \mathbf{Z})$ are evaluated in [10] for general degree n . For a regular elliptic conjugacy classes which can be represented by M in $Sp(3, \mathbf{Z})$ and $\mathrm{diag}[\lambda_1, \lambda_2, \lambda_3]$ in $U(3)$, respectively; its contribution is given by

$$N_{\{M\}} = |C_{M,Z}|^{-1}(\bar\lambda_1\bar\lambda_2\bar\lambda_3)^k/(1-\bar\lambda_1^2)(1-\bar\lambda_2^2)(1-\bar\lambda_3^2)(1-\bar\lambda_1\bar\lambda_2)(1-\bar\lambda_2\bar\lambda_3)(1-\bar\lambda_2\bar\lambda_3)$$

when $k > 4$. Here $C_{M,Z}$ is the contralizer of M in $Sp(3, \mathbf{Z})$ and $|C_{M,Z}|$ is its order as a subgroup of $Sp(3, \mathbf{Z})/\{\pm 1\}$.

A direct calculation by TABLE II , we get the following.

THEOREM 1. The total contribution from conjugacy classes of regular elliptic elements in $Sp(3, Z)$ to the dimension formula is the sum of K_i $(i = 1, 2, \dots, 24)$ given as follow:

(1) $K_1 = 2^{-12}3^{-1}[1,-1]$, (No. 1 in TABLE II) ,

(2) $K_2 = 2^{-6}$, (No. 2 and 3 in TABLE II) ,

(3) $K_3 = 2^{-3}3^{-1}[1,-1]$, (No. 4 in TABLE II) ,

(4) $K_4 = -2^{-4}3^{-7}$, (No. 5 and 9 in TABLE II) ,

(5) $K_5 = 2^{-5}3^{-5}[1,-2,1]$, (No. 6 and 8 in TABLE II) ,

(6) $K_6 = 2^{-2}3^{-5}$, (No. 7 and 10 in TABLE II),

(7) $K_7 = 2^{-1}3^{-4}$ (No. 11 and 12 in TABLE II),

(8) $K_8 = 2^{-3}3^{-3}[-1,2,-1]$, (No. 13 and 15 in TABLE II) ,

(9) $K_9 = 2^{-1}3^{-3}$ (No. 14 and 16 in TABLE II) ,

(10) $K_{10} = 3^{-2}[1,0,1,0,-1,0,0,-1,0]$,

(No. 17, 18, 73-78 in TABLE II) ,

(11) $K_{11} = 2^{-7}3^{-1}[-1,2,-1]$, (No. 19 and 20 in TABLE II) ,

(12) $K_{12} = 2^{-4}3^{-1}[1,2,1,-1,-2,-1]$,

(No. 23-26 in TABLE II) ,

(13) $K_{13} = 2^{-3}3^{-3}[1,2,1,-1,-2,-1]$,

(No. 27-30 in TABLE II) ,

(14) $K_{14} = 2^{-3}3^{-3}[1,-1]$, (No. 31 and 32 in TABLE II) ,

(15) $K_{15} = 2^{-4}3^{-3}[1,2,1,-1,-2,-1]$,

(No. 33 and 34 in TABLE II),

(16) $K_{16} = 2^{-4}3^{-2}[1,2,1,-1,-2,-1]$,

(No. 35 and 37 in TABLE II) ,

(17) $K_{17} = 2^{-4}3^{-2}[1,2,1,-1,-2,-1]$,

(No. 36 and 38 in TABLE II) ,

(18) $K_{18} = 2^{-6}$, (No. 39 and 40 in TABLE II) ,

(19) $K_{19} = 2^{-3}3^{-2}[1,-1]$, (No. 41 and 42 in TABLE II) ,

(20) $K_{20} = 2^{-1}5^{-1}[1,0,1,1,-1,-1,0,-1,-1,1]$,

(No. 43-50 in TABLE II) ,

(21) $K_{21} = 2^{-4}3^{-1}[1,-2,1]$, (No. 51-54 in TABLE II) ,

(22) $K_{22} = 2^{-4}3^{-4}[1,-2,1]$, (No. 51 and 56 in TABLE II) ,

(23) $K_{23} = 3^{-1}5^{-1}[1,0,1,0,0,-1,0,0,0,0,0,0,-1,0,0]$,

(No. 57-72 in TABLE II) ,

(24) $K_{24} = 7^{-1}[1,0,1,0,0,0,0]$,

(No.79-86 in TABLE II) .

REMARK. Besides K_{20}, K_{23} and K_{24}, the remaining K's are periodic rational-valued functions in k of period 1 or 2 or 6 or 12.

For further convenience, we add certain K's together and have the following.

COROLLARY 1. *The total contribution from conjugacy classes of regular elliptic elements in* $Sp(3, \mathbf{Z})$ *to the dimension formula is given by*

$$N_0 = 7^{-1}[1,0,1,0,0,0,0] + 2^{-2}5^{-1}[1,0,1,1,-1,-1,0,-1,-1,1]$$
$$+ 3^{-1}5^{-1}[1,0,1,0,0,-1,0,0,0,0,0,0,-1,0,0]$$
$$+ 3^{-2}[1,0,1,0,-1,0,0,-1,0] + (-2^{-4}3^{-7} + 2^{-6} + 2^{-2}3^{-5}\cdot 25)$$
$$+ (2^{-12}3^{-1} + 2^{-6}3^{-3}\cdot 131)[1,-1]$$
$$+ (-2^{-7}3^{-1} + 2^{-5}3^{-5}\cdot 133)[1,-2,1]$$
$$+ 2^{-4}3^{-3}\cdot 17[1,2,1,-1,-2,-1] .$$

3.3 Contributions from Conjugacy Classes in Γ_3^2 .

Conjugacy classes of Γ_3^2 are given explicitly in TABLE III except those with trivial contribution and their contributions can be computed by Theorem 2 and Theorem 3 in 4.3 (page 95 and 97) of [11]. For torsion element which can be represented by $E_2 \times M$ in $Sp(3, Z)$ and $\mathrm{diag}[1, \lambda_1, \lambda_2]$ in $U(3)$; the contribution from the conjugacy class represented by $E_2 \times M$ to the dimension formula is given by

$$N_{\{M\}}$$
$$= |C_{M,Z}|^{-1}\cdot 2^{-4}3^{-1}(\bar{\lambda}_1\bar{\lambda}_2)^k[(2k-4)/(1-\bar{\lambda}_1^2)(1-\bar{\lambda}_1\bar{\lambda}_2)(1-\bar{\lambda}_2^2)(1-\bar{\lambda}_1)(1-\bar{\lambda}_2)$$
$$+ 2/(1-\bar{\lambda}_1^2)(1-\bar{\lambda}_2^2)(1-\bar{\lambda}_1)^2(1-\bar{\lambda}_2)^2] .$$

Here $C_{M,Z}$ is the centralizer of M in $Sp(2, \mathbf{Z})$ and $|C_{M,\mathbf{Z}}|$ is the order of $C_{M,Z}$ as a subgroup of $Sp(2, \mathbf{Z})/\{\pm 1\}$.

For the family of conjugacy classes represented by

$\begin{bmatrix} 1 & s \\ 0 & 1 \end{bmatrix} \in M$, $s \in \mathbf{Z} - \{0\}$, their total contribution is given by

$$N'_{\{M\}} = |C_{M,\mathbf{Z}}|^{-1}(-2^{-2})(\bar{\lambda}_1\bar{\lambda}_2)^k/(1-\bar{\lambda}_1^2)(1-\bar{\lambda}_1\bar{\lambda}_2)(1-\bar{\lambda}_2^2)(1-\bar{\lambda}_1)(1-\bar{\lambda}_2).$$

THEOREM 2. The total contribution from conjugacy classes in Γ_3^2 is the sum of K_i $(i = 25, \ldots, 36)$ given as follow:

(1) $K_{25} = -2^{-11}3^{-1} - 2^{-10}3^{-1}$ (1-1 ~ 1-10 in TABLE III),

(2) $K_{26} = -2^{-6}3^{-4}(2k-4)[1,0,-1] + 2^{-5}3^{-4}[1,-2,1] + 2^{-4}3^{-3}[1,0,-1]$,
(2-1 ~ 2-4 in TABLE III),

(3) $K_{27} = 2^{-6}3^{-5}(2k-4)[1,0,-1] - 2^{-5}3^{-4}[1,-2,1] - 2^{-2}3^{-4}[1,0,-1]$,
(3-1 ~ 3-8 in TABLE III),

(4) $K_{28} = -2^{-6}3^{-1}[1,-1]$, (4-1 ~ 4-8 in TABLE III),

(5) $K_{29} = -2^{-6}3^{-1}$, (5-1 ~ 5-8 in TABLE III),

(6) $K_{30} = 2^{-3}3^{-3}$, (6-1 ~ 6-6 in TABLE III),

(7) $K_{31} = 2^{-3}3^{-4}$, (7-1 ~ 7-6 in TABLE III),

(8) $K_{32} = 2^{-6}3^{-3}(2k-4)[1,0,-1] - 2^{-5}3^{-4}\cdot 17[1,-2,1] - 2^{-5}3^{-1}[1,0,-1]$,
(8-1 ~ 8-3 and 9-1 ~ 9-3 in TABLE III),

(9) $K_{33} = 2^{-5}3^{-4}(2k-4)[1,0,-1] - 2^{-4}3^{-5}\cdot 17[1,-2,1] - 2^{-4}3^{-2}[1,0,-1]$,
(10-1 ~ 10-3 and 11-1 ~ 11-3 in

TABLE III),

(11) $K_{35} = 2^{-6}3^{-2}(2k-4)[1,0,-1,-1,0,1] - 2^{-2}3^{-1}[1,0,-1,-1,0,1]$
$- 2^{-3}3^{-2}[1,2,1,-1,-2,-1]$,
(13-1 ~ 13-8, 14-1 ~ 14-8, 15-1
15-8 and 16-1 ~ 16-8 in TABLE III)

(12) $K_{36} = 2^{-3}3^{-1}5^{-2}(2k-4)[1,0,-1,3,-3]$
$+ 2^{-3}3^{-1}5^{-2}[-66,0,54,-54,66]$
(17-1 ~ 17-24 in TABLE III),

Proof. Here we only prove (11), (12) will be proved in next chapter and the remaining follow by the same procedures.

Firstly, we calculate the contribution from conjugacy classes represented by 13-1, 14-1, 15-1 and 16-1. Their total contribution is

$$A_1 = \frac{1}{12}\cdot 2^{-4}3^{-1}\{(2k-4)\,\Sigma\,(\bar\lambda_1\bar\lambda_2)^k/(1-\bar\lambda_1^2)(1-\bar\lambda_1\bar\lambda_2)(1-\bar\lambda_2^2)(1-\bar\lambda_1)(1-\bar\lambda_2)$$

$$+ 2\,\Sigma\,(\bar\lambda_1\bar\lambda_2)^k/(1-\bar\lambda_1^2)(1-\bar\lambda_2^2)(1-\bar\lambda_1)^2(1-\bar\lambda_2)^2 \quad .$$

In the summation, $\lambda_1 = i$, $\lambda_2 = \rho$, ρ^2, ρ^4 and ρ^5 with $\rho = e^{\pi i/3}$. It follows

$$A_1 = \frac{1}{12}\cdot 2^{-4}3^{-1}(2k-4)\cdot [1,0,-1,-1,0,1]$$

$$= 2^{-6}3^{-2}(2k-4)[1,0,-1,-1,0,1] \ .$$

Secondly, the contribution from families of conjugacy classes represented by 13-2, 14-2, 15-2 and 16-2 is

$$A_2 = \frac{1}{12} \cdot (-2\bar{\lambda}^2)\ \Sigma\ (\bar{\lambda}_1\bar{\lambda}_2)^k/(1-\bar{\lambda}_1^2)(1-\bar{\lambda}_1\bar{\lambda}_2)(1-\bar{\lambda}_2^2)(1-\bar{\lambda}_1)(1-\bar{\lambda}_2)$$

$$= -2^{-4}3^{-1}[1,0,-1,-1,0,1] .$$

To compute the contribution from conjugacy classes represented by 13-3, 14-3, 15-7 and 16-7, we need the follow property of Hurwitz-zeta function.

$$\lim_{\varepsilon\to 0} \left(\frac{1}{i(s - 1/3)}\right)^{1+\varepsilon} = -\pi + \pi i/\sqrt{3} .$$

The total contribution from these families of conjugacy classes is

$$A_3 = \frac{1}{12}\ 2^{-1} \cdot \mathrm{Trace}_{Q(i,\sqrt{3})/Q}[(-i\bar{\rho}^2)^k/(1-\bar{\rho}^4)(1-\bar{\rho}^2)(1+i)(1-i^2)(1+i\bar{\rho}^2)$$

$$\times (-1 + i/\sqrt{3}\ )]$$

$$= -2^{-4}3^{-1}[1,0,-1,-1,0,1] + 2^{-4}3^{-2}[1,2,1,-1,-2,-1] .$$

Note that we also have

$$\lim_{\varepsilon\to 0} \left(\frac{1}{i(s - 1/2)}\right)^{1+\varepsilon} = -\pi$$

and

$$\lim_{\varepsilon\to 0} \left(\frac{1}{i(s - 5/6)}\right)^{1+\varepsilon} = -\pi - \sqrt{3}\,\pi i .$$

Suppose that the total contributions from families of conjugacy classes represented by 13-4, 13-8, 14-4, 14-8, 15-3, 15-6, 16-3, 16-6 and 13-5, 14-5, 15-8, 16-8, are A_4 and A_5, respectively. Then

$$A_4 = -2^{-4}3^{-1}[1,0,-1,-1,0,1]$$

and

$$A_5 = -2^{-4}3^{-1}[1,0,-1,-1,0,1] - 2^{-4}3^{-1}[1,2,1,-1,-2,-1] .$$

Now it follows

$$\begin{aligned} K_{35} &= A_1 + A_2 + A_3 + A_4 + A_5 \\ &= 2^{-6}e^{-2}(2k-4)[1,0,-1,-1,0,1] - 2^{-4}3^{-1}[1,0,-1,-1,0,1] \\ &\quad - 2^{-3}3^{-2}[1,2,1,-1,-2,-1] . \end{aligned}$$

This proves our assertion in (11).

COROLLARY 2. The total contribution from conjugacy classes in Γ_3^2 to the dimension formula is

$$\begin{aligned} N_1 &= 2^{-3}3^{-1}5^{-2}(2k-4)[1,0,-1,3,-3] + 2^{-3}3^{-1}5^{-2}[-66,0,54,-54,66] \\ &\quad + 2^{-6}3^{-2}(2k-4)[1,0,-1,-1,0,1] - 2^{-2}3^{-1}[1,0,-1,-1,0,1] \\ &\quad - 2^{-3}3^{-2}[1,2,1,-1,-2,-1] \\ &\quad - 2^{-6}3^{-5}\cdot 13(2k-4)[1,0,-1] + (-2^{-11}-2^{-4}3^{-4}\cdot 19) \\ &\quad + 2^{-5}3^{-4}\cdot 61[1,0,-1] - 2^{-5}3^{-5}\cdot 112[1,-2,1] . \end{aligned}$$

3.4 Contributions from conjugacy classes in Γ_3^1.

Conjugacy classes of Γ_3^1 are shown in TABLE IV and they can be divided into following kinds by their characteristic polynomials.

(A) Characteristic polynomial = $(X-1)^4(X^2+1)$ or $(X-1)^4(X^2+1)$
(1-1 ~ 1-5 in TABLE IV),

(B) Characteristic polynomial = $(X-1)^4(X^2+X+1)$ or $(X-1)^4(X^2-X+1)$
(2-1 ~ 2-3 and 3-1 ~ 3-7 in TABLE IV),

(C) Characteristic polynomial = $(X^2-1)^2(X^2+1)$
(1-6 ~ 1-22 in TABLE IV),

(D) Characteristic polynomial = $(X^2-1)(X^2+X+1)$ or $(X^2-1)^2(X^2-X+1)$
(2-8 ~ 2-27 in TABLE IV),

(E) Characteristic polynomial = $(X^2+1)^3$
(1-23 ~ 1-28 in TABLE IV),

(F) Characteristic polynomial = $(X^2+X+1)^3$
(2-32 ~ 2-35 in TABLE IV),

(G) Characteristic polynomial = $(X^2+X+1)(X^2+1)^2$
(2-4 ~ 2-7 in TABLE IV),

(H) Characteristic polynomial = $(X^2+1)(X^2+X+1)^2$
(1-6 and 1-7 in TABLE IV),

(I) Characteristic polynomial = $(X^2-X+1)(X^2+X+1)^2$
(2-28 and 2-37 in TABLE IV).

Contributions from these conjugacy classes can be computed by formulae obtained in 4.4, 4.5, 4.6 and 4.7 of [11].

THEOREM 3. *Contributions from conjugacy classes in* Γ_3^1 *to the dimension formula are* K_i $(i = 32, \ldots, 45)$ *given as follow*: ($\phi(X)$ represents characteristic polynomial of conjugacy classes.)

(A) $K_{37} = -2^{-12}3^{-2}5^{-1}(2k-4)^2[1,-1] + 2^{-8}3^{-1}[1,-1],$

$$(\phi(X) = (X-1)^4(X^2+1) \quad \text{or} \quad (X+1)^4(X^2+1)).$$

(B) $K_{38} = 2^{-9}3^{-5}5^{-1}(2k-3)(2k-4)(2k-5)[-2,0,2]$
$- 2^{-9}3^{-5}5^{-1}(2k-4)(2k-5)[10,-20,10]$
$+ 2^{-9}3^{-5}5^{-1}(2k-4)[9,10,-18] - 2^{-8}3^{-4}\cdot 24(2k-4)[1,0,-1]$
$+ 2^{-5}3^{-3}[1,-2,1] + 2^{-5}3^{-4}\cdot 7[1,0,-1],$

$$(\phi(X) = (X-1)^4(X^2+X+1) \quad \text{or} \quad (X-1)^4(X^2-X+1)),$$

(C) $K_{39} = [2^{-12}3^{-2}\cdot 7(2k-4)^2 - 2^{-11}3^{-1}\cdot 16(2k-4) + 2^{-10}\cdot 12]$
$\times [1,-1],$

$$(\phi(X) = (X^2-1)^2(X^2+1)),$$

(D) $K_{40} = 2^{-9}3^{-4}\cdot 7(2k-4)(2k-6)[1,-2,1] + 2^{-8}3^{-4}\cdot 7(2k-4)[0,-2,2]$
$- 2^{-8}3^{-4}\cdot 24(2k-4)[1,0,-1] - 2^{-8}3^{-4}\cdot 48(2k-4)[1,-2,1]$
$+ 2^{-4}3^{-2}[1,0,-1] + 2^{-4}3^{-3}\cdot 5[1,-2,1],$

$$(\phi(X) = (X^2-1)^2(X^2+X+1)),$$

(E) $K_{41} = [2^{-12}3^{-2}\cdot 6(2k-3)(2k-5) - 2^{-8}3^{-1}\cdot 4(2k-4)$
$+ 2^{-12}3^{-1} + 2^{-7}\cdot 13] \times [1,-1],$

$$(\phi(X) = (X^2+1)^3),$$

(F) $K_{42} = 2^{-4}3^{-6}(2k-3)(2k-5)[1,-2,1] - 2^{-4}3^{-6}(2k-3)[1,0,-1]$
$- 2^{-2}3^{-7}[2,-1,-1] - 2^{-4}3^{-3}(2k-4)[1,-2,1]$
$+ 2^{-2}3^{-1}[1,-2,1] - 2^{-3}3^{-5}[1,0,-1],$
$(\phi(X) = (X^2+X+1)^3),$

(G) $K_{43} = 2^{-7}3^{-2}\cdot 5(2k-4)[1,0,-1] - 2^{-7}3^{-2}\cdot 5[1,-2,1]$
$- 2^{-3}3^{-1}[1,0,-1],$
$(\phi(X) = (X^2+X+1)(X^2+1)^2),$

(H) $K_{44} = -2^{-2}3^{-3}[1,-1], \quad (\phi(X) = (X^2+1)(X^2+X+1)^2),$

(I) $K_{45} = 2^{-3}3^{-4}(2k-4)[1,0,-1] + 2^{-4}3^{-4}[1,-2,1]$
$- 2^{-1}3^{-2}[1,0,-1],$
$(\phi(X) = (X^2-X+1)(X^2+X+1)^2),$

Proof. Here we shall only prove (C) and (G). Apply Theorem 20 in 4.5 of [11] to conjugacy classes from 2-8 to 2-11 of TABLE IV, we get the following contributions

(2- 8) $2^{-9}3^{-4}(2k-4)(2k-6)[1,-2,1] + 2^{-8}3^{-4}(2k-4)[0,2,-2],$

(2- 9) $-2^{-8}3^{-3}(2k-4)[1,-2,1] - 2^{-8}3^{-3}[1,0,-1],$

(2-10) $-2^{-8}3^{-3}(2k-4)[1,-2,1] + 2^{-8}3^{-2}[1,0,-1],$

(2-11) $2^{-6}3^{-2}[1,-2,1].$

On the other hand, contributions from 2-12 to 2-15 are given as follow:

(2-12) $2^{-8}3^{-3}(2k-4)(2k-6)[1,-2,1] + 2^{-7}3^{-3}(2k-4)[0,2,-2],$

(2-13) $-2^{-8}3^{-2}(2k-4)[1,-2,1] - 2^{-8}3^{-2}[1,0,-1],$

(2-14) $-2^{-8}3^{-2}(2k-4)[1,-2,1] + 2^{-8}3^{-1}[1,0,-1]$,

(2-15) $2^{-6}3^{-2}[1,-2,1]$.

Contributions from 2-16 to 2-27 are computed by the same formula and the properties of Hurwitz-zeta function as used in the previous Theorem. Sum all resulted contributions together, we get K_{40}.

Apply Theorem 21 and 22 in 4.6 of [11] to conjugacy classes 2-4 and 2-5 in TABLE IV, we get the following:

(2- 4) $2^{-6}3^{-2}(2k-4)[1,0,-1] - 2^{-6}3^{-2}[1,-2,1]$,

(2- 5) $-2^{-4}3^{-1}[1,0,-1]$.

By the same arguement as used in p.35 of [11], we get contributions from conjugacy classes 2-5 and 2-6 are

(2- 6) $2^{-7}3^{-1}(2k-4)[1,0,-1] - 2^{-7}3^{-1}[1,-2,1]$,

(2- 7) $-2^{-4}3^{-1}[1,0,-1]$.

Add (2-4), (2-5), (2-6) and (2-7) together, we get K_{43}.

3.5 Contributions from Conjugacy Classes in Γ_3^0.

Conjugacy classes of Γ_3^0 are given explicitly in Section 2.5 and their contributions can be computed by Theorems in Chapter V of [11]. Especially, contributions from conjugacy classes in the principal congruence subgroups $\Gamma_3(2)$ and $\Gamma_3(N)$ $(N \geqslant 3)$ of $Sp(3, Z)$ are calculated in [11].

THEOREM 4. The contributions from conjugacy classes in $\Gamma_3^0 - (\Gamma_3^1 \cup \Gamma_3^2)$ to the dimension formula are K_i (i = 46, ..., 51) given as follow:

(1) $K_{46} = 2^{-15}3^{-6}5^{-2}7^{-1}(2k-2)(2k-3)(2k-4)^2(2k-5)(2k-6)$
$- 2^{-9}3^{-2}5^{-1}(2k-4) + 2^{-7}3^{-3}$,

(Characteristic polynomial = $(X-1)^6$, (1) of Lemma 5 in 2.5)

(2) $K_{47} = 2^{-15}3^{-4}5^{-1} \cdot 31(2k-2)(2k-4)^2(2k-6)$
$- 2^{-13}3^{-3}5^{-1} \cdot 16(2k-3)(2k-4)(2k-5)$
$- 2^{-11}3^{-3} \cdot 16(2k-3)(2k-5) + 2^{-9}3^{-2} \cdot 6(2k-4)$
$+ 2^{-10}3^{-2} \cdot 7(2k-4) - 2^{-8}3^{-1} \cdot 2$,

(Characteristic polynomial = $(X-1)^4(X+1)^2$, (2) of Lemma 5 in 2.5)

(3) $K_{48} = 2^{-5}3^{-4}(2k-3)(2k-5) - 2^{-4}3^{-3} \cdot 4(2k-4) + 2^{-2}3^{-1}$,

(Characteristic polynomial = $(X-1)^2(X^2-X+1)^2$, (4) of Lemma 5 in 2.5)

(4) $K_{49} = 2^{-5}3^{-5} \cdot 13(2k-3)(2k-5) - 2^{-4}3^{-4} \cdot 16(2k-4)$
$+ 2^{-2}3^{-2} + 2^{-1}3^{-2}$

(Characteristic polynomial = $(X+1)^2(X^2-X+1)^2$, (5) of Lemma 5 in 2.5)

(5) $K_{50} = 2^{-11}3^{-2} \cdot 23(2k-3)(2k-5) - 2^{-9}3^{-1} \cdot 51(2k-4) + 2^{-6} \cdot 7$,

(Characteristic polynomial = $(X-1)^2(X^2+1)^2$, (3) of Lemma 5 in 2.5)

(6) $K_{51} = 2^{-10}3^{-1}(2k-4) - 2^{-7}3^{-1}$,

(Characteristic polynomial = $(X-1)^4(X+1)^2$, (6) of Lemma 5 in 2.5).

Proof. Here we only prove (4).

For $s_1 = s_2 = 0$ in (4) of Lemma 5 of 2.5, the contribution is

$$\frac{1}{6}\cdot\frac{2^{-8}\pi^{-3}(2k-3)(2k-5)}{(1-\cos 2\theta)(1-\cos\theta)}\bigg|_{\theta=\pi/3}\frac{1}{3}\cdot\frac{4\pi^3}{3}$$

$$= 2^{-5}3^{-4}(2k-3)(2k-5)$$

by Theorem 14 of Chapter 5 of [11]. If $s_2 = 0$ and s_1 ranges over all nonzero integers, the contribution is

$$\frac{1}{6}\cdot\frac{2^{-6}\pi^{-2}(2k-4)}{(1-\cos 2\theta)(1-\cos\theta)}\bigg|_{\theta=\pi/3}\lim_{\varepsilon\to 0}\sum_{s_1\neq 0}\left(\frac{1}{is_1}\right)^{1+\varepsilon}\cdot\frac{4\pi}{3}$$

$$= -2^{-4}3^{-4}(2k-4).$$

Suppose $s_1 = 0$ and s_2 ranges over all nonzero integers. The contribution is

$$\frac{1}{6}\cdot\frac{2^{-6}\pi^{-2}(2k-4)}{(1-\cos 2\theta)(1-\cos\theta)}\bigg|_{\theta=\pi/3}\frac{\pi}{3}\cdot\lim_{\varepsilon\to 0}\sum_{s_2\neq 0}\left(\frac{3}{is_2}\right)^{1+2\varepsilon}\cdot 4$$

$$= -2^{-4}3^{-3}(2k-4).$$

Finally, as s_1 and s_2 both range over all nonzero integers, the contribution is

$$\frac{1}{6}\cdot\frac{2^{-5}\pi^{-2}}{(1-\cos 2\theta)(1-\cos\theta)}\bigg|_{\theta=\pi/3}(-\pi)\cdot(-12\pi)$$

$$= 2^{-2}3^{-1}.$$

Add these contributions together, we get K_{48}.

REMARK. Contributions from conjugacy classes in Γ_3^0 are polynomial functions in the weight k.

3.6 An Explicit Dimension Formula for Siegel Cusp Forms of Degree Three.

To get an explicit dimension formula for Siegel cusp forms of degree three with respect to $Sp(3, \mathbf{Z})$, we can simply sum all K_i $(i = 1, \ldots, 52)$ in Theorem 1, 2, 3 and 4 together. However, we want to get a formula which can be easily applied when it is needed.

To do so, we first observe that all contributions can be written as a finite sum of products of polynomials and periodic rational-values functions in the weight k. Thus we have

$$\dim_{\mathbf{C}} S(k; Sp(3, \mathbf{Z})) = \sum_{\gamma} C_\gamma(k) P_\gamma(k) ,$$

where γ ranges over all representatives of conjugacy classes of toision elements in $Sp(3, \mathbf{Z})$ with possible repetition, $C_\gamma(k)$ is a periodic rational valued function in k and $P_\gamma(k)$ is a polynomial in k.

Among K_i or terms in K_i $(i = 1, \ldots, 51)$, we shall extract those terms with at least one of the following properties:

(1) the period of $C_\gamma(k)$ is not a divisor of 12,

(2) the degeee of $P_\gamma(k)$ is greater than 2,

(3) the denominator of $C_\gamma(k) P_\gamma(k)$ has a divisor 5 or 7.

These terms are list as follow:

(1) $L_1 = 2^{-15}3^{-6}5^{-2}7^{-1}(2k-2)(2k-3)(2k-5)^2(2k-6)$, the contribution from identity E_6 ;

(2) $L_2 = 2^{-15}3^{-4}5^{-1} \cdot 31(2k-2)(2k-4)^2(2k-6)$, the contribution from two conjugacy classes of torsion elements of order 2 with characteristic polynomial $(X-1)^4(X+1)^2$;

(3) $L_3 = -2^{-13}3^{-3}5^{-1} \cdot 16(2k-3)(2k-4)(2k-5)$, the contribution from two families of conjugacy classes represented by $E_4 \times \begin{bmatrix} -1 & s \\ 0 & -1 \end{bmatrix}$ $(s \neq 0)$ in $Sp(3, \mathbf{R})$;

(4) $L_4 = 2^{10}3^{-5}5^{-1}(2k-3)(2k-4)(2k-5)[-2,0,2]+2^{-10}3^{-5}5^{-1}(2k-4)(2k-5)[-10,20,-10]+2^{-9}3^{-5}5^{-1}(2k-4)[8,10,-18]$, the contribution from conjugacy classes of torsion elements of order 3 or 6 with characteristic polynomial $(X-1)^4(X^2+X+1)$ or $(X-1)^4(X^2-X+1)$;

(5) $L_5 = -2^{-12}3^{-2}5^{-1}(2k-4)[1,-1]$, the contribution from conjugacy classes of torsion elements of order 4 with characteristic polynomial $(X-1)^4(X^2+1)$;

(6) $L_6 = -2^{-9}3^{-2}5^{-1}(2k-4)$, the contribution from conjugacy classes represented by $[S, E_3] \in Sp(3, \mathbf{Z})$, rank $S = 2$;

(7) $L_7 = 2^{-3}3^{-1}5^{-1}(2k-4)[1,0,-1,3,-3]+2^{-2}3^{-1}5^{-2}[-66,0,54,-54,66]$, the contribution from conjugacy classes of elements with characteristic polynomial $(X-1)^2(X^4+X^3+X^2+X+1)$ or $(X+1)^2(X^4+X^3+X^2+X+1)$;

(8) $L_8 = \frac{1}{7}[1,0,1,0,0,0,0]$, the contribution from elements of order 7;

(9) $L_9 = \frac{1}{9}[1,0,1,0,-1,0,0,-1,0]$, the contribution from elements of order 9;

(10) $L_{10} = \frac{1}{20}[1,0,1,1,-1,-1,0,-1,-1,1]$, the contribution from elements of order 20;

(11) $L_{11} = \frac{1}{15}[1,0,1,0,0,-1,0,0,0,0,0,0,-1,0,0]$, the contribution from elements of order 15 or 30.

The remaining terms can be combined together as

$$C_1(k)(2k-4)^2 + C_2(k)(2k-4) + C_3(k)$$

with $C_j(k)$ $(j = 1,2,3)$ rational-valued function in k of period 12; i.e. $C_j(k+12) = C_j(k)$. By Theorem 1, 2, 3 and 4, we get

$$C_1(k) = 2^{-7}3^{-2}[4,2,4,3,3,3]+2^{-12}3^{-6}[451,1249,451,937,763,937],$$
$$C_2(k) = -2^{-3}3^{-1}+2^{-8}3^{-6}[-3010,783,-4496,-1714,-1161,-1904],$$
$$C_3(k) = 2^{-4}3^{-6}[5314,0,8770,2560,2916,2128].$$

<u>MAIN THEOREM 1</u>. <u>For even integer</u> $k \geqslant 10$, <u>the dimension formula for the vector space of Siegel cusp forms of degree three and weight</u> k <u>is given by</u>

$$\dim_{\mathbf{C}} S(k; Sp(3, \mathbf{Z})) = \sum_{i=1}^{11} L_i + C_1(k)(2k-4)^2 + C_2(k)(2k-4) + C_3(k)$$

<u>with</u> L_i <u>as shown previously and</u>

$$C_1(k) = 2^{-7}3^{-2}[4,2,4,3,3,3]+2^{-12}3^{-6}[451,1249,451,937,763,937],$$
$$C_2(k) = -2^{-3}3^{-1}+2^{-8}3^{-6}[-3010,783,-4496,-1714,-1161,-1904],$$
$$C_3(k) = 2^{-4}3^{-6}[5314,0,8770,2560,2916,2128].$$

Here is a table shows explicit values of $\dim_{\mathbf{C}} S(k; Sp(3, \mathbf{Z}))$ when $10 \leqslant k \leqslant 118$

k	10	12	14	16	18	20	22	24	26	28	30
$\dim_{\mathbf{C}} S(k; Sp(3,\mathbf{Z}))$	0	1	1	3	4	6	9	14	17	27	34

32	34	36	38	40	42	44	46	48	50	52	54	56	58	60
46	61	82	99	135	165	208	261	325	389	490	584	708	852	1023

62	64	66	68	70	72	74	76	78	80	82
1200	1445	1687	1984	2327	2717	3133	3663	4199	4838	5557

84	86	88	90	92	94	96	98	100	102
6360	7225	8267	9344	10585	11968	13489	15116	17037	19023

104	106	108	110	112	114	116	118
21271	23742	26429	29324	32615	36050	39881	44047

REMARK. It is easy to see that

$$C_2(k) = -2^{-8}3^{-4}\cdot 1087 - 2^{-7}[1,-1]+2^{-6}3^{-2}[1,0,-1,-1,0,1]$$
$$-2^{-3}3^{-3}[1,-2,1]+2^{-8}3^{-6}\cdot 419[1,0,-1]$$

and

$$C_3(k) = 2^{-4}3^{-7}\cdot 10844+2^{-3}3^{-3}\cdot 19[1,-1]+2^{-4}3^{-3}[1,2,1,-1,-2,-1]$$
$$+2^{-4}3^{-7}\cdot 3235[1,-2,1]+2^{-4}3^{-4}\cdot 84[-1,0,1].$$

3.7 Automorphic Forms of Degree Three and Its Generating Function.

When k is an even integer, it is well known [20] that the generating of degree two is

$$\frac{1}{(1-T^4)(1-T^6)(1-T^{10})(1-T^{12})}.$$

In other words if we let $A_n(k)$ be the vector space of automorphic forms defined on H_n of degree n and wieght k, then

$$\sum_{k=0}^{\infty} \dim_{\mathbf{C}} A_2(k) T^k = \frac{1}{(1-T^4)(1-T^6)(1-T^{10})(1-T^{12})}$$

as a formal power series.

Note that the Siegel Φ-operator (boundary operator) defined by

$$\phi(f)(Z) = \lim_{t \to \infty} f \begin{bmatrix} Z & 0 \\ 0 & it \end{bmatrix}$$

is an onto mapping from $A_n(k)$ to $A_{n-1}(k)$ when $k > n+1$. Thus, in particular, we have

$$\dim_{\mathbf{C}} A_3(k) = \dim_{\mathbf{C}} A_2(k) + \dim_{\mathbf{C}} S(k; Sp(3,\mathbf{Z})), \ k \geq 10, \ k \text{ even}.$$

Also we know that (For example, see p.50 of Siegelsche Modulfunktionen by E. Freitag.)

$$\dim_{\mathbf{C}} S(k; Sp(3,\mathbf{Z})) = 0 \quad \text{if} \quad 0 \leq k \leq 8.$$

For $k = 8$, all integrals and zeta functions involving in our computation are still convergent. Thus our formula is still valid for $k = 8$, this yields $\dim_{\mathbf{C}} S(8; Sp(3,\mathbf{Z})) = 0$. Hence we have

$$\begin{aligned}\sum_{k=0}^{\infty} \dim_{\mathbf{C}} A_3(k) T^k = {} & 1 + T^4 + T^6 + T^{10} + 4T^{12} + 3T^{14} + 7T^{16} \\ & + 8T^{18} + 11T^{20} + 15T^{22} + 22T^{24} + 24T^{26} \\ & + 37T^{28} + 45T^{30} + 58T^{32} + 75T^{34} + 99T^{36} \\ & + 115T^{38} + 156T^{40} + 187T^{42} + 232T^{44} \\ & + 288T^{46} + 356T^{48} + 460T^{50} + 527T^{52}\end{aligned}$$

$$+ 623T^{54} + 750T^{56} + 898T^{58} + 1075T^{60} + 1252T^{62} + 1505T^{64} + 1750T^{66} + 2051T^{68} + 2400T^{70} + 2797T^{72} + (\text{Higher terms})$$

A direct calculation of residues shows

$$\sum_{k=0}^{\infty} \dim_{\mathbb{C}} A_3(k) T^k = P(T)/(1-T^4)(1-T^{12})^2(1-T^{14})(1-T^{18})(1-T^{20})(1-T^{30})$$

with

$$\begin{aligned} P(T) = {} & 1 + T^6 + T^{10} + T^{12} + 3T^{16} + 2T^{18} + 2T^{20} + 5T^{22} \\ & + 4T^{24} + 5T^{26} + 7T^{28} + 6T^{30} + 9T^{32} + 10T^{34} + 10T^{36} \\ & + 12T^{38} + 14T^{40} + 15T^{42} + 16T^{44} + 18T^{46} + 18T^{48} \\ & + 19T^{50} + 21T^{52} + 19T^{54} + 21T^{56} + 21T^{58} + 19T^{60} \\ & + 21T^{62} + 19T^{64} + 18T^{66} + 18T^{68} + 16T^{70} + 15T^{72} \\ & + 14T^{74} + 12T^{76} + 10T^{78} + 10T^{80} + 9T^{82} + 6T^{84} + 7T^{86} \\ & + 5T^{88} + 4T^{90} + 5T^{92} + 2T^{94} + 2T^{96} + 3T^{98} + T^{102} \\ & + T^{104} + T^{108} + T^{114}. \end{aligned}$$

This result is consistent with those by Prof. Tsugumine [35] by a different method. However, we have difficulty to verify his proof.

REMARK. Our formula for $\dim_{\mathbb{C}} S(k; Sp(3, \mathbf{Z}))$ given in Main Theorem I is true even for $k = 2$ and 6, but it is not true for $k = 0$ and $k = 4$.

CHAPTER IV

DIMENSION FORMULAE FOR THE VECTOR SPACES OF SIEGEL CUSP FORMS OF DEGREE THREE

4.1 Introduction

Dimension formulae for modular forms of degree two is known in the early works of Igusa [20] in 1966. After that, several mathematicans obtained the dimension formula with respect to the principal congruence subgroup $\Gamma_2(N)$ $(N \geqslant 3)$ of $Sp(2, \mathbf{Z})$ via Selberg Trace Formula [8, 9, 25, 29] or Riemann-Roch Theorem [34, 36].

In 1981, the author successfully obtained the dimension of $S(k; Sp(2, \mathbf{Z}))$ and presented an effective procedure for the computation of all the terms necessary in the determination of $\dim_{\mathbf{C}} S(k; Sp(3, \mathbf{Z}))$ [10, 11]. In particular, we have

$$\dim_{\mathbf{C}} S(k; Sp(3, \mathbf{Z})) = \sum_{\gamma} C_{\gamma}(k) P_{\gamma}(k)$$

where γ ranges over all conjugacy classes of torsion elements in $Sp(3, \mathbf{Z})$, $C_{\gamma}(k)$ is a period function in k depending also on eigenvalues of γ and $P_{\gamma}(k)$ is a polynomial in k with degree no larger than the complex dimension of fixed subvariety of γ on H_3.

To find the explicit values of $\dim_{\mathbf{C}} S(k; Sp(3, \mathbf{Z}))$, it remains to

(1) determine conjugacy classes of elements whose characteristic polynomials are products of cyclotomic polynomials,

(2) calculate contributions from conjugacy classes in (1) by Theorems in [11] or [21].

Conjugacy classes of $Sp(3, \mathbf{Z})$ are given in CHAPTER I and CHAPTER II as a preparation to our evaluation of $\dim_{\mathbf{C}} S(k; Sp(3, \mathbf{Z}))$. However, it needs a long computation to carry out the second step since the number of conjugacy classes in $Sp(3, \mathbf{Z})$ is very large. Note that even a small mistake in the course of calculation of contributions may lead to an incorrect dimension formula. So in this chapter, we shall devolop a method to know the contributions will be before the computation and reprove the following theorems.

MAIN THEOREM I. <u>For even integer</u> $k \geqslant 10$, <u>the dimension formula for the vector space of Siegel cusp forms of degree three and weight</u> k <u>is given by</u>

$$\dim_{\mathbf{C}} S(k, Sp(3, \mathbf{Z})) = \text{Sum of Main Terms} + C_1(k)(2k-4)^2 + C_2(k)(2k-4) + C_3(k)$$

<u>where the main terms and the values of</u> C_j $(j = 1,2,3)$ <u>are given by</u> TABLE V <u>as follow:</u>

TABLE V. Main Terms in the Dimension Formula

No.	Contribution	Conjugacy Classes
1	$2^{-15}3^{-6}5^{-2}7^{-1}(2k-2)(2k-3)(2k-4)^2(2k-5)(2k-6)$	E_6
2	$2^{-15}3^{-4}5^{-1}31(2k-2)(2k-4)^2(2k-6)$	$[1, 1, -1]$
3	$-2^{-13}3^{-3}5^{-1}16(2k-3)(2k-4)(2k-5)$	$E_4 \times \begin{bmatrix} -1 & s \\ 0 & -1 \end{bmatrix}$, $(s \neq 0)$

TABLE V (CONTINUED)

No.	Contribution	Conjugacy Classes
4	$2^{-10}3^{-5}5^{-1}(2k-3)(2k-4)(2k-5) \times [-2,\ 0,\ 2]$	$[1,\ 1,\ e^{i\theta}]$
5	$2^{-10}3^{-5}5^{-1}(2k-4)(2k-5) \times [-10,\ 20,\ -10]$	$\theta = \frac{\pi}{3},\ \frac{2\pi}{3},\ \frac{4\pi}{3},\ \frac{5\pi}{3}$
6	$2^{-9}3^{-5}5^{-1}(2k-4) \times [8,\ 10,\ -18]$	
7	$-(-1)^{k/2}2^{-12}3^{-2}5^{-1}(2k-4)^2$	$[1,\ 1,\ \pm i]$
8	$-2^{-9}3^{-2}5^{-1}(2k-4)$	$[S,\ E_3]$, rank $S = 2$
9	$2^{-3}3^{-1}5^{-2}(2k-4) \times [1,\ 0,\ -1,\ 3,\ -3]$	Elements with characteristic polynomial
10	$2^{-3}3^{-1}5^{-2} \times [-66,\ 0,\ 54,\ -54,\ 66]$	$(X\pm1)^2(X^4\pm X^3+X^2\pm X+1)$
11	$\frac{1}{7}\ [1,\ 0,\ 1,\ 0,\ 0,\ 0,\ 0]$	Elements of order 7
12	$\frac{1}{9}\ [1,\ 0,\ 1,\ 0,\ -1,\ 0,\ 0,\ -1,\ 0]$	Elements of order 9
13	$\frac{1}{20}\ [1,\ 0,\ 1,\ 1,\ -1,\ -1,\ 0,\ -1,\ -1,\ 1]$	Elements of order 20
14	$\frac{1}{15}\ [1,0,1,0,0,-1,0,0,0,0,0,0,-1,0,0]$	Elements of order 30

15 The remaining term is $C_1(k)(2k-4)^2 + C_2(k)(2k-4) + C_3(k)$, where

$$C_1(k) = 2^{-7}3^{-2}[4,\ 2,\ 4,\ 3,\ 3,\ 3\ ] + 2^{-12}3^{-6}[451,\ 1249,\ 451,\ 937,\ 763,\ 937],$$

$$C_2(k) = -2^{-3}3^{-1} + 2^{-8}3^{-6}\ [-3010,\ 783,\ -4496,\ -1714,\ -1161,\ -1904],$$

$$C_3(k) = 2^{-4}3^{-6}[5314,\ 0,\ 8770,\ 2560,\ 2916,\ 2128]$$

* Here $C(k) = [a_0,\ a_1,\ \ldots,\ a_{m-1}]$ means $C(k) = a_j$ if $k \equiv 2j \pmod{2m}$ for $0 \leqslant j \leqslant m-1$.

MAIN THEOREM II. <u>The dimension formula for the vector space of Siegel cusp forms of degree three with respect to the congruence subgroup</u> $\Gamma_3(2)$ <u>of</u> $\Gamma_3 = Sp(3,\ \mathbf{Z})$ <u>is given by</u>

$$\begin{aligned}&\dim_{\mathbf{C}} S(k,\ \Gamma_3(2)) \\ &= [\Gamma_3 : \Gamma_3(2)] \times [\,2^{-15}3^{-6}5^{-2}7^{-1}(2k-2)(2k-3)(2k-4)^2(2k-5)(2k-6) \\ &\quad + 2^{-15}3^{-4}5^{-1}(2k-2)(2k-4)^2(2k-6) \\ &\quad - 2^{-14}3^{-3}5^{-1}(2k-3)(2k-4)(2k-5) - 2^{-13}3^{-3}(2k-3)(2k-5)\end{aligned}$$

$$- 2^{-14}3^{-2}5^{-1}(2k-4) + 2^{-13}3^{-1}(2k-4) - 2^{-13}3^{-1} + 2^{-13}3^{-3}]$$

for an even integer $k \geqslant 10$, where $[\Gamma_3 \colon \Gamma_3(2)] = 2^9 3^4 .35$.

MAIN THEOREM III. The dimension formula for the vector space of siegel cusp forms of degree three with respect to the congruence subgroup $\Gamma(N)$ $(n \geqslant 3)$ of $\Gamma_3 = Sp(3, \mathbf{Z})$ is given by

$$\begin{aligned} &\dim_{\mathbf{C}} S(k;\ \Gamma_3(N)) \\ &= [\Gamma_3 \colon \Gamma_3(N)] \times [2^{-15}3^{-6}5^{-2}7^{-1}(2k-2)(2k-3)(2k-4)(2k-5)(2k-6) \\ &\quad - 2^{-9}2^{-2}5^{-1}(2k-4)N^{-5} + 2^{-7}3^{-3}N^{-6}], \end{aligned}$$

where k is an even integer greater than 9 and

$$[\Gamma_3 \colon \Gamma_3(N)] = \frac{1}{2}N^{21} \sum_{\substack{p|N \\ p:\ \text{prime}}} (1-p^{-2})(1-p^{-4})(1-p^{-6}).$$

The main ideal in our compatation is to extract certain contributions which we call main terms so that the sum of remaining terms appear to be the form

$$C_1(k)(2k-4)^2 + C_2(k)(2k-4) + C_3(k)$$

with $C_j(k+12) = C_j(k)$, $j = 1, 2, 3$. Note that C_j $(j = 1,2,3)$ must satisfy certain equations so that the sum of main terms and remaining terms is an integer. With the help of this observation, we can determine values of C_j $(j = 1,2,3)$ accurately by a long computation.

The method we employed here applies to cases of higher degrees. Indeed, we did reduce the problem of finding

$\dim_{\mathbf{C}} S(k; Sp(n, \mathbf{Z}))$, at least for the case $n = 1, 2, 3$; to the problem of

(1) finding conjugacy classes of regular elliptic elements in $Sp(n, \mathbf{Z})$,

(2) calculating contributions from certain conjugacy classes.

and

(3) determining values of certain constants.

The problem in (1) is treated in [24, 32]. Thus we can write down conjugacy classes of elements whose characteristic polynomials are products of cyclotomic polynomials by an induction on the degree n. The problem in (2) is treated in [21] is a more general context. The problem in (3) can be treated by our knowledge of modular forms of lower weights instead of direct computation. In our determination of $\dim_{\mathbf{C}} S(k; Sp(3, \mathbf{Z}))$, the constants C_j $(j = 1,2,3)$ can be determine uniquely by $\dim_{\mathbf{C}} S(k; Sp(3, \mathbf{Z}))$ when $10 \leqslant k \leqslant 44$ and the sum of main terms as shown in TABLE V.

4.2 Eie's Results.

Results in this section are based on the computation carried out in [11].

A conjugacy classes $\{M\}$ of the element M in $Sp(3, \mathbf{Z})$ has a possible nonzero contribution to the dimension formula only when

(1) M is an element of finite order

or

(2) M is conjugate in $Sp(3, \mathbf{R})$ to an element of the form

$M' \cdot [S, E_3]$ with M' is an element of finite order which has a positive dimensional fixed subvariety on H_3.

In (2), we call M' and $[S, E_3]$ are torsion part and parabolic part of M respectively.

For the first case, we let Ω denote the subvariety of fixed points of M. Then the contribution is given by N_i $(i = 0, 1, 2, 3, 4, 5, 6, 7, 8)$ as follows:

(a) $\dim_{\mathbf{C}} \Omega = 6$, $M = E_6$, the identity;

$$N_0 = 2^{-15}3^{-6}5^{-2}7^{-1}(2k-2)(2k-3)(2k-4)^2(2k-5)(2k-6) .$$

(Theorem 3, 5.3; CHAPTER V of [11])

(b) $\dim_{\mathbf{C}} \Omega = 4$, M is conjugate in $Sp(3, \mathbf{R})$ to $[1, 1, -1]$;

$$N_1 = c \cdot 2^{-15}3^{-4}5^{-1}(2k-2)(2k-4)^2(2k-6).$$

($c = 1$ if $M = [1, 1, -1]$; Theorem 4, 5.3; CHAPTER V of [11])

(c) $\dim_{\mathbf{C}} \Omega = 3$, $M = [1, 1, \lambda]$;

$$N_2 = \frac{c \cdot \bar{\lambda}^k (2k-4)}{2^{10}2^3 5} \left\{ \frac{(2k-3)(2k-5)}{(1-\bar{\lambda})^3 (1+\bar{\lambda})} + \frac{3(2k-5)}{(1-\bar{\lambda})^4} + \frac{6}{(1-\bar{\lambda})^5} \right\}.$$

(Theorem 4, 4.4; CHAPTER IV of [11])

(d) $\dim_{\mathbf{C}} \Omega = 2$, $M = [1, -1, \lambda]$;

$$N_3 = c \cdot 2^{-9}3^{-2}\bar{\lambda}^k \left\{ \frac{(2k-4)(2k-6)}{(1-\bar{\lambda}^2)^2} + \frac{4(2k-4)}{(1-\bar{\lambda}^2)^3} \right\}.$$

(Theorem 20, 4.5; CHAPTER IV of [11])

(e) $\dim_{\mathbf{C}} \Omega = 2$, $M = [\lambda, \lambda, \bar{\lambda}]$;

$$N_4 = c \cdot 2^{-3}\bar{\lambda}^k \left\{ \frac{(2k-3)(2k-5)}{(1-\bar{\lambda}^2)^3(1-\lambda^2)} + \frac{(2k-3)(\lambda^4-1)}{(1-\bar{\lambda}^2)^3(1-\lambda^2)^2} + \frac{2}{(1-\bar{\lambda}^2)^3(1-\lambda^2)^3} \right\},$$

(Theorem 23, 4.7; CHAPTER IV of [11])

(f) $\dim_{\mathbf{C}} \Omega = 2$, $M = [1, \lambda, \bar{\lambda}]$;

$$N_5 = c \cdot \frac{2^{-6}(2k-3)(2k-5)}{(1-\bar{\lambda})(1-\lambda)(1-\bar{\lambda}^2)(1-\lambda^2)} .$$

(Theorem 14, 5.5; CHAPTER V of [11])

(g) $\dim_{\mathbf{C}} \Omega = 1$, $M = [1, \lambda_1, \lambda_2]$;

$$N_6 = \frac{c \cdot 2^{-4}3^{-1}(\bar{\lambda}_1\bar{\lambda}_2)^k}{(1-\bar{\lambda}_1^2)(1-\bar{\lambda}_1\bar{\lambda}_2)(1-\bar{\lambda}_2^2)} \left\{ \frac{2k-4}{(1-\bar{\lambda}_1)(1-\bar{\lambda}_2)} + \frac{2(1-\bar{\lambda}_1\bar{\lambda}_2)}{(1-\bar{\lambda}_1)^2(1-\bar{\lambda}_2)^2} \right\} .$$

(Theorem 2, 4.3; CHAPTER IV of [11])

(h) $\dim_{\mathbf{C}} \Omega = 1$, $M = [\lambda_1, \lambda_2, \bar{\lambda}_2]$;

$$N_7 = \frac{c \cdot 2^{-2}\bar{\lambda}_1^k}{(1-\bar{\lambda}_1^2)(1-\bar{\lambda}_2^2)(1-\lambda_2^2)} \left\{ \frac{2k-4}{(1-\bar{\lambda}_1\bar{\lambda}_2)(1-\bar{\lambda}_1\lambda_2)} + \frac{1-\bar{\lambda}_1^2}{(1-\bar{\lambda}_1\lambda_2)^2(1-\bar{\lambda}_1\lambda_2)^2} \right\}.$$

(i) $\dim_{\mathbf{C}} \Omega = 0$, $M = [\lambda_1, \lambda_2, \lambda_3]$;

$$N_8 = |C_{M,\mathbf{Z}}|^{-1}(\bar{\lambda}_1\bar{\lambda}_2\bar{\lambda}_3)^k \prod_{1\le i\le j\le 3} (1 - \bar{\lambda}_i\bar{\lambda}_j)^{-1} .$$

(Theorem 1, 4.2; CHAPTER IV of [11])

In the above formulae, $\lambda, \lambda_1, \lambda_2, \lambda_3$ denote roots of unity which are different from ± 1 and satisfy $\lambda_i\lambda_j \neq 1$ for all i,j, and c is a rational number which depends only on $\mathrm{vol}(C_{M,\mathbf{Z}}\backslash C_{M,\mathbf{R}})$ if we choose the representative M in $Sp(3, \mathbf{Z})$.

For the second case, we have to choose a suitable family of conjugacy classes so that the total contribution is a number of one of the types (a) to (i).

Here are some typical examples which appear in our calculations.

(j) $M = E_4 \times \begin{bmatrix} -1 & s \\ 0 & -1 \end{bmatrix}$. As s runs over all nunzero integers, the total contribution is

$$-2^{-13}3^{-3}5^{-1}(2k-3)(2k-4)(2k-5).$$

(Theorem 7, 5.4; CHAPTER V of [11])

(k) $M = \begin{bmatrix} 1 & s \\ 0 & 1 \end{bmatrix} \times [0, U]$ with $U = [1, -1]$. As s runs over all nonzero integers, the total contribution is

$$-2^{-11}3^{-3}(2k-3)(2k-5).$$

(Theorem 11, 5.4; CHAPTER V of [11])

(ℓ) $M = \begin{bmatrix} a & b \\ c & d \end{bmatrix} \times \begin{bmatrix} E & S \\ 0 & E \end{bmatrix}$, $\begin{bmatrix} a & b \\ c & d \end{bmatrix}$ being conjugate in $SL_2(\mathbf{R})$ to $\begin{bmatrix} \cos\theta & \sin\theta \\ -\sin\theta & \cos\theta \end{bmatrix}$, $\lambda = e^{i\theta}$ $(\sin\theta \neq 0)$. The contribution is

(1) $$\frac{-2^{-7}3^{-2}\bar{\lambda}^{k}}{|G|}\left\{\frac{2k-4}{(1-\bar{\lambda})^3(1+\bar{\lambda})} + \frac{1}{(1-\bar{\lambda})^4}\right\} \quad \text{if rank } S = 1,$$

(2) $\dfrac{2^{-6}3^{-1}\bar{\lambda}^{k}}{|G|} \cdot \dfrac{1}{(1-\bar{\lambda})^{3}(1+\bar{\lambda})}$ if rank $S = 2$.

Here M is an element of $Sp(3, \mathbf{Z})$ and G is the centralizer of $\begin{bmatrix} a & b \\ c & d \end{bmatrix}$ in $SL_2(\mathbf{Z})$.

(Theorem 11, 4.3; CHAPTER IV of [11])

(m) $M = \begin{bmatrix} a & b \\ c & d \end{bmatrix} \times \begin{bmatrix} 1 & s_1 \\ 0 & 1 \end{bmatrix} \times \begin{bmatrix} -1 & s_2 \\ 0 & -1 \end{bmatrix}$ with $\begin{bmatrix} a & b \\ c & d \end{bmatrix}$ and G as in the previous case. The contribution of conjugacy classes represented by such M's to the dimension formula is:

(1) $\dfrac{-2^{-7}3^{-1}\bar{\lambda}^{k}}{|G|} \left\{ \dfrac{2k-4}{(1-\bar{\lambda}^{2})^{2}} + \dfrac{1}{(1-\bar{\lambda})^{3}(1+\bar{\lambda})} \right\}$ if $s_2 = 0$ and s_1 run over all nonzero integers.

(2) $\dfrac{-2^{-7}3^{-1}\bar{\lambda}^{k}}{|G|} \left\{ \dfrac{2k-4}{(1-\bar{\lambda}^{2})^{2}} + \dfrac{1}{(1-\bar{\lambda})(1+\bar{\lambda})^{3}} \right\}$ if $s_1 = 0$ and s_2 run over all nonzero integers.

(3) $\dfrac{2^{-5-k}}{|G|(1-\bar{\lambda}^{2})^{2}}$ if s_1, s_2 run over all nonzero integers.

(Theorem 20, 4.5; CHAPTER IV of [11])

(n) $M = \begin{bmatrix} 1 & s \\ 0 & 1 \end{bmatrix} \times \begin{bmatrix} P & Q \\ R & S \end{bmatrix}$ with $\begin{bmatrix} P & Q \\ R & S \end{bmatrix}$ is conjugate in $Sp(2, \mathbf{R})$ to $\lambda_1, \lambda_2, \lambda_1^2, \lambda_1\lambda_2, \lambda_2^2 \neq 1$, and has contralizer G in $Sp(2, \mathbf{Z})$. The total contribution as s runs over all nonzero integers is

$$\frac{-2^{-2}(\bar{\lambda}_1\bar{\lambda}_2)^k}{|G|(1-\bar{\lambda}_1^2)(1-\bar{\lambda}_1\bar{\lambda}_2)(1-\bar{\lambda}_2^2)(1-\bar{\lambda}_1)(1-\bar{\lambda}_2)} .$$

(Theorem 3, 4.3; CHAPTER IV of [11]).

After excluding contribution with zero contribution as shown in [11], we now can write the dimension formula as a finite sum as follows.

PROPOSITION 1. *For even integer* $k \geqslant 10$, *we have*

$$\dim_{\mathbf{C}} S(k;\ Sp(3,\ \mathbf{Z})) = \sum_{\gamma} C_\gamma(k) P_\gamma(k),$$

where γ *ranges over all representatives of conjugacy classes of torsion elements in* $Sp(3,\ \mathbf{Z})$, $C_\gamma(k)$ *is a period function in* k *depending only on eigenvalues of* γ *and* $P_\gamma(k)$ *is a polynomial function in* k *with degree no larger that the complex dimension of fixed subvariety of* γ.

REMARK: In the real calculation, we combine certain contributions so that values of $C_\gamma(k)$ are rational numbers. For example, if we combine contributions from conjugacy classes represented by $[1, \pm i, \rho]$, $[1, \pm i, \rho^2]$, $[1, \pm i, \rho^4]$ and $[1, \pm i, \rho^5]$. The total contribution is

$$2^{-6}3^{-2} \times (2k-4) \times [1,\ 0,\ -1,\ -1,\ 0,\ 1].$$

Here we may choose $C(k) = 2^{-6}3^{-2} \times [1,\ 0,\ -1,\ -1,\ 0,\ 1]$ and $P(k) = (2k-4)$.

4.3 Conjugacy Classes of $Sp(3, \mathbf{Z})$

Conjugacy classes of $Sp(3, \mathbf{Z})$ are given; at least those conjugacy classes with possible nonzero contribution to the dimension formula, explicitly in Chapter I and II.

Here we only give representatives of conjugacy classes which contribute to the main terms in our formula.

1. The identity E_6 forms a conjugacy class with characteristic polynomial $(X-1)^6$.

2. There are two conjugacy classes of order 2 with characteristic polynomial $(X-1)^4(X+1)^2$, i.e. $[1, 1, -1]$ and $\delta^{-1}[1, 1, -1]\delta$ with $\delta = [T, E_3]$, where

$$T = \begin{bmatrix} 0 & 0 & 0 \\ 0 & 0 & \frac{1}{2} \\ 0 & \frac{1}{2} & 0 \end{bmatrix}.$$

3. There are two families of conjugacy classes of elements which are conjugacy in $Sp(3, \mathbf{R})$ to $E_4 \times \begin{bmatrix} -1 & s \\ 0 & -1 \end{bmatrix}$, $s \neq 0$; i.e.

$$\mathcal{E} = \left\{ E_4 \times \begin{bmatrix} -1 & s \\ 0 & -1 \end{bmatrix} \mid s \in \mathbf{Z} - \{0\} \right\}$$

and $\delta^{-1}\mathcal{E}\,\delta$.

4. $E_4 \times \begin{bmatrix} 0 & 1 \\ -1 & 0 \end{bmatrix}$ and $E_4 \times \begin{bmatrix} 0 & -1 \\ 1 & 0 \end{bmatrix}$ are representatives of conjugacy classes of torsion elements of order four with characteristic $(X-1)^4(X^2+1)$.

5. The representatives of torsion elements of order 3 or 6 with characteristic polynomial $(X-1)^2(X^2-X+1)$ are

$$E_4 \times \begin{bmatrix} 0 & -1 \\ 1 & -1 \end{bmatrix}, \quad E_4 \times \begin{bmatrix} -1 & 1 \\ -1 & 0 \end{bmatrix}, \quad E_4 \times \begin{bmatrix} 1 & -1 \\ 1 & 0 \end{bmatrix} \text{ and }$$

$$E_4 \times \begin{bmatrix} 0 & 1 \\ -1 & 1 \end{bmatrix}.$$

6. $[S, E_3]$, (rank $S = 2$, $S = {}^tS \in M_3(\mathbf{Z})$) are representatives of conjugacy classes of elements characterized by those M in $Sp(3, \mathbf{Z})$ with rank $(M - E_6) = 2$ and the characteristic polynomal of M is $(X-1)^6$.

7. Conjugacy classes of elements with characteristic polynomial $(X-1)^2(X^4+X^3+X^2+X+1)$ or $(X+1)^2(X^4+X^3+X^2+X+1)$ are given in the following :

$$M = \begin{bmatrix} -1 & 0 & 1 & 0 \\ 0 & 0 & 0 & 1 \\ 1 & 0 & 0 & -1 \\ 1 & -1 & -1 & 0 \end{bmatrix}.$$

7-1 $[M; 0, 0; 0] \sim e[2/5, 4/5, 0]$.

7-2 $[-M; 0, 0; 0] \sim e[2/5, 4/5, 1]$.

7-3 $[M^2; 0, 0; 0] \sim e[4/5, 8/5, 0]$.

7-4 $[-M^2; 0, 0; 0] \sim e[4/5, 8/5, 1]$.

7-5 $[M^3; 0, 0; 0] \sim e[6/5, 2/5, 0]$.

7-6 $[-M^3; 0, 0; 0] \sim e[6/5, 2/5, 1]$.

7-7 $[M^4; 0, 0; 0] \sim e[8/5, 6/5, 0]$.

7-8 $[-M^4; 0, 0; 0] \sim e[8/5, 6/5, 1]$.

7-9 $[M; 0, 0; s]$, $s \in \mathbf{Z} - \{0\}$.

7-10 $[M; {}^t[1,0], 0; s] \sim [M_{17}; 0, 0; s-\frac{2}{5}]$, $s \in \mathbf{Z}$.

7-11 $[M;\ {}^t[2,0],\ 0;\ s] \sim [M_{17};\ 0,\ 0;\ s-\frac{8}{5}],\quad s \in \mathbf{Z}$.

7-12 $[-M;\ 0,\ 0;\ s],\quad s \in \mathbf{Z} - \{0\}$.

7-13 $[M^2;\ 0,\ 0;\ s],\quad s \in \mathbf{Z} - \{0\}$.

7-14 $[M^2;\ {}^t[1,0],\ 0;\ s] \sim [M^2;\ 0,\ 0;\ s+\frac{2}{5}],\quad s \in \mathbf{Z}$.

7-15 $[M^2;\ {}^t[2,0],\ 0;\ s] \sim [M^2;\ 0,\ 0;\ s+\frac{8}{5}],\quad s \in \mathbf{Z}$.

7-16 $[-M^2;\ 0,\ 0;\ s],\ s \in \mathbf{Z} - \{0\}$.

7-17 $[M^3;\ 0,\ 0;\ s],\quad s \in \mathbf{Z} - \{0\}$.

7-18 $[M^3;\ {}^t[1,0],\ 0;\ s] \sim [M^3;\ 0,\ 0;\ s+\frac{3}{5}],\quad s \in \mathbf{Z}$.

7-19 $[M^3;\ {}^t[2,0],\ 0;\ s] \sim [M^3;\ 0,\ 0;\ s+\frac{12}{5}],\quad s \in \mathbf{Z}$.

7-20 $[-M^3;\ 0,\ 0;\ s],\quad s \in \mathbf{Z} - \{0\}$.

7-21 $[M^4;\ 0,\ 0;\ s],\ s \in \mathbf{Z} - \{0\}$.

7-22 $[M^4;\ {}^t[1,0],\ 0;\ s] \sim [M^4;\ 0,\ 0;\ s+\frac{2}{5}],\ s \in \mathbf{Z}$.

7-23 $[M^4;\ {}^t[2,0],\ 0;\ s] \sim [M^4;\ 0,\ 0;\ s+\frac{8}{5}],\ s \in \mathbf{Z}$.

7-24 $[-M^4;\ 0,\ 0;\ s],\ s \in \mathbf{Z} - \{0\}$.

* Here we use the following notation from [13].

$$[M;\ p,\ q;\ s] = \begin{bmatrix} P & 0 & Q & 0 \\ 0 & 1 & 0 & 0 \\ R & 0 & S & 0 \\ 0 & 0 & 0 & 1 \end{bmatrix} \begin{bmatrix} E & 0 & 0 & q \\ {}^tp & 1 & {}^tq & {}^tpq+s \\ 0 & 0 & E & -p \\ 0 & 0 & 0 & 1 \end{bmatrix},\ M = \begin{bmatrix} P & Q \\ R & S \end{bmatrix} .$$

8. Conjugacy classes of elements of order 7, 9, 20 and 30 are list as follow:

TABLE VI Conjugacy Classes of Elements of Order 7, 9, 20 and 30.

No	Representative in U(3)	Characteristic polynomial	Order of Centralizer
1	e[1/2, 2/5, 4/5]	$(X^2+1)P_1(X)$	20
2	e[1/2, 4/5, 8/5]	$(X^2+1)P_1(X)$	"
3	e[1/2, 6/5, 2/5]	$(X^2+1)P_1(X)$	"
4	e[1/2, 8/5, 6/5]	$(X^2+1)P_1(X)$	"
5	e[3/2, 2/5, 4/5]	$(X^2+1)P_1(-X)$	"
6	e[3/2, 4/5, 8/5]	$(X^2+1)P_1(-X)$	"
7	e[3/2, 6/5, 2/5]	$(X^2+1)P_1(-X)$	"
8	e[3/2, 8/5, 6/5]	$(X^2+1)P_1(-X)$	"
9	e[1/3, 2/5, 4/5]	$(X^2-X+1)P_1(X)$	30
10	e[2/3, 2/5, 4/5]	$(X^2-X+1)P_1(X)$	"
11	e[4/3, 2/5, 4/5]	$(X^2+X+1)P_1(x)$	"
12	e[5/3, 2/5, 4/5]	$(X^2+X+1)P_1(X)$	"
13	e[1/3, 4/5, 8/5]	$(X^2-X+1)P_1(X)$	"
14	e[2/3, 4/5, 8/5]	$(X^2-X+1)P_1(X)$	"
15	e[4/3, 4/5, 8/5]	$(X^2+X+1)P_1(X)$	"
16	e[5/3, 4/5, 8/5]	$(X^2+X+1)P_1(X)$	"
17	e[1/3, 6/5, 2/5]	$(X^2-X+1)P_1(X)$	"
18	e[2/3, 6/5, 2/5]	$(X^2-X+1)P_1(X)$	"
19	e[4/3, 6/5, 2/5]	$(X^2+X+1)P_1(X)$	"
20	e[5/3, 6/5, 2/5]	$(X^2+X+1)P_1(X)$	"
21	e[1/3, 8/5, 6/5]	$(X^2-X+1)P_1(X)$	"
22	e[2/3, 8/5, 6/5]	$(X^2-X+1)P_1(X)$	"
23	e[4/3, 8/5, 6/5]	$(X^2+X+1)P_1(X)$	"
24	e[5/3, 8/5, 6/5]	$(X^2+X+1)P_1(X)$	"
25	e[2/9, 4/9, 8/9]	X^6+X^3+1	9
26	e[4/9, 8/9, 16/9]	X^6+X^3+1	"
27	e[8/9, 16/9, 14/9]	X^6+X^3+1	"

TABLE VI (Continued)

No	Representative in U(3)	Characteristic polynomial	Order of Centralizer
28	e[10/9, 2/9, 4/9]	X^6+X^3+1	9
29	e[14/9, 10/9, 2/9]	X^6+X^3+1	"
30	e[16/9, 14/9, 10/9]	X^6+X^3+1	"
31	e[2/9, 8/9, 14/9]	X^6+X^3+1	"
32	e[4/9, 16/9, 10/9]	X^6+X^3+1	"
33	e[2/7, 4/7, 6/7]	$P_2(X)$	7
34	e[4/7, 8/7, 12/7]	$P_2(X)$	"
35	e[6/7, 12/7, 4/7]	$P_2(X)$	"
36	e[8/7, 2/7, 10/7]	$P_2(X)$	"
37	e[10/7, 6/7, 2/7]	$P_2(X)$	"
38	e[12/7, 10/7, 8/7]	$P_2(X)$	"
39	e[2/7, 4/7, 8/7]	$P_2(X)$	"
40	e[6/7,12/7,10/7]	$P_2(X)$	"

* Here $P_1(X) = X^4+X^3+X^2+X+1$ and $P_2(X) = X^6+X^5+X^4+X^3+X^2+X+1$.

4.4 The Main Terms

By Eie's results as shown in Section 4.2, we can write

$$\dim_{\mathbf{C}} S(k;\ Sp(3,\ \mathbf{Z})) = \sum_{\gamma} C_\gamma(k) P_\gamma(k).$$

Now we shall extract those terms with the property

(1) the period of $C_\gamma(k)$ is not a divisor of 12,

(2) the degree of $P(k)$ is greater than 2

or (3) the denominator of value $C(k)P(k)$ has a divisor 5 or 7.

These terms are precisely contributions from conjugacy classes shown in Section 2. Contributions from conjugacy classes of regular elliptic elements of order 7, 9, 20 and 30 are easily calculated by (i) in Section 1. Here we only discuss the remaining cases.

PROPOSITIONS 2. *The contribution from identity* E_6 *to the dimension formula is*

$$2^{-15}3^{-6}5^{-2}7^{-1}(2k-2)(2k-3)(2k-4)^2(2k-5)(2k-6).$$

Proof. This follows from the fact that the volume of the fundamental domain of H_3 with respect to $Sp(3, \mathbf{Z})$ is $3^{-6}5^{-2}7^{-1}\pi^6$, and the contribution is given by

$$2^{-15}\pi^{-6}(2k-2)(2k-3)(2k-4)^2(2k-5)(2k-6) \times 3^{-6}5^{-2}7^{-1}\pi^6$$

$$= 2^{-15}3^{-6}5^{-2}7^{-1}(2k-2)(2k-3)(2k-4)^2(2k-5)(2k-6).$$

PROPOSITIONS 3. *The contribution of elements in* $Sp(3, \mathbf{Z})$ *which are conjugate in* $Sp(3, \mathbf{Z})$ *to* $[1, 1, -1]$ *or* $\delta^{-1}[1, 1, -1]\delta$ *as in (2) of Section 4.3, to the dimension formula is*

$$2^{-15}3^{-4}5^{-1} \times 31 \times (2k-2)(2k-4)^2(2k-6).$$

Proof. The contribution from the conjugacy class represented by $[1, 1, -1]$ is $2^{-15}3^{-4}5^{-1}(2k-2)(2k-4)^2(2k-6)$ by Eie's result [11]. Now it suffices to compute the contribution from the conjugacy classes represented by $\delta^{-1}[1, 1, -1]\delta$.

Let

$$G = \{M \in Sp(2, \mathbf{R}) \times SL_2(R) \mid \delta^{-1}M\delta \in Sp(3, \mathbf{Z})\}.$$

The $\delta^{-1}G\delta$ is the centralizer of $\delta^{-1}[1, 1, -1]\delta$ in $Sp(3, \mathbf{Z})$ and a direct verification shows $G \subset Sp(2, \mathbf{Z}/2) \times SL_2(\mathbf{Z}/2)$ Furthermore, we have

$$\mathrm{vol}(G \backslash Sp(2,\mathbf{R}) \times SL_2(R)) = 30 \cdot \frac{\pi^3}{270} \cdot \frac{\pi}{3}.$$

Note that $[1, 1, -1]$ and $\delta^{-1}[1, 1, -1]\delta$ are conjugate in $Sp(3, \mathbf{R})$. Hence the contribution of $\delta^{-1}[1, 1, -1]\delta$ is

$$2^{-15}3^{-4}5^{-1}(2k-2)(2k-4)^2(2k-6) \times 30.$$

This proves our assertion.

In the same way, we prove

PROPOSITION 4. <u>The contribution of elements in</u> $Sp(3, \mathbf{Z})$ <u>which are conjugate in</u> $Sp(3, \mathbf{Z})$ <u>to elements of</u> $\mathfrak{E}$ <u>or</u> $\delta^{-1}\mathfrak{E}\delta$ <u>as in 3 of Section 4.3 to the dimension formula is</u>

$$-2^{-13}3^{-3}5^{-1} \times 16 \times (2k-3)(2k-4)(2k-5).$$

Applying Theorem 12 of Chapter 4 in [11] to the case $|G| = 2$, $\theta = \pm\frac{\pi}{2}$ and $|G| = 3$, $\theta = \pm\frac{\pi}{3}, \pm\frac{2\pi}{3}$, we obtain the follow:

PROPOSITION 5. *The contribution of elements in* $Sp(3, \mathbb{Z})$ *which are conjugate in* $Sp(3, \mathbb{Z})$ *to*

$$E_4 \times \begin{bmatrix} 0 & 1 \\ -1 & 0 \end{bmatrix} \quad \text{or} \quad E_4 \times \begin{bmatrix} 0 & -1 \\ 1 & 0 \end{bmatrix}$$

to the dimension formula is

$$(-1)^{k/2} 2^{-12} 3^{-2} 5^{-1} (2k-4)^2.$$

PROPOSITION 6. *The contribution of elements in* $Sp(3, \mathbb{Z})$ *which are conjugate in* $Sp(3, \mathbb{Z})$ *to*

$$E_4 \times \begin{bmatrix} 0 & -1 \\ 1 & -1 \end{bmatrix} \quad \text{or} \quad E_4 \times \begin{bmatrix} 1 & -1 \\ 1 & 0 \end{bmatrix} \quad \text{or}$$

$$E_4 \times \begin{bmatrix} -1 & 1 \\ -1 & 0 \end{bmatrix} \quad \text{or} \quad E_4 \times \begin{bmatrix} 0 & 1 \\ -1 & 1 \end{bmatrix}$$

to the dimension formula is given by

$$2^{-10} 3^{-5} 5^{-1} (2k-3)(2k-4)(2k-5) \times [-2, 0, 2]$$

$$+2^{-10} 3^{-5} 5^{-1} (2k-4)(2k-5) \times [-10, 20, -10]$$

$$+2^{-9} 3^{-5} 5^{-1} (2k-4) \times [8, 10, -18].$$

Here $\alpha(k) = [a, b, c]$ *stands for*

$$\alpha(k) = \begin{cases} a & \text{if } k \equiv 0 \pmod 6, \\ b & \text{if } k \equiv 2 \pmod 6, \\ c & \text{if } k \equiv 4 \pmod 6. \end{cases}$$

In general, when there is no ambiguity, we shall let $\alpha(k) = [a_0, a_1, \ldots, a_{m-1}]$ to stand for the function $\alpha(k) = a_j$ when

$k \equiv 2j \pmod{2m}$.

PROPOSITION 7. *The contribution of elements in* $Sp(3, \mathbf{Z})$ *which are conjugate in* $Sp(3, \mathbf{Z})$ *to* $[S, E_3]$, $S = {}^tS \in M_3(\mathbf{Z})$, *rank* $S = 2$ *to the dimension formula is*

$$-2^{-9}3^{-2}5^{-1}(2k-4).$$

Proof. This is precisely Theorem 6 in CHAPTER 5 of [11].

PROPOSITION 8. *The contribution of elements in* $Sp(3, \mathbf{Z})$ *with characteristic polynomial* $(X-1)^2(X^4+X^3+X^2+X+1)$ *or* $(X-1)^2(X^4-X^3+X^2-X+1)$ *to the dimension formula, is given by*

$$2^{-3}3^{-1}5^{-2} \times (2k-4) \times [1, 0, -1, 3, -3]$$
$$+2^{-3}3^{-1}5^{-2} \times [-66, 0, 54, 66].$$

Proof. The contribution from elements of order 5 and 10 are

$$2^{-4}3^{-1}5^{-2}(2k-4) \times [2,0,-2,1,-1] + 2^{-3}3^{-1}5^{-2} \times [-1,0,-1,1,1]$$

and

$$2^{-4}3^{-1}5^{-1}(2k-4) \times [0,0,0,1,-1] + 2^{-3}3^{-1}5^{-1} \times [-1,0,-1,1,1]$$

respectively by a direct calculation by (h) in Section 1. Also the contribution from families of conjugacy classes with $s \in \mathbf{Z} - \{0\}$ (7-9, 7-12, 7-13, 7-16, 7-17, 7-20, 7-21, 7-24) is

$$-2^{-2}5^{-2} \times [2,\ 0,\ -2,\ 1,\ -1]$$

$$-2^{-2}5^{-1} \times [0,\ 0,\ 0,\ 1,\ -1]\ .$$

The contribution from conjugacy classes represented by $[M;\ 0,\ 0;\ s - \frac{2}{5}]$ and $\{M;\ 0,\ 0;\ s - \frac{8}{5}\}$ is given by

$$N = \frac{1}{10} \cdot \frac{\zeta^{-6k_\pi - 1}}{(1-\zeta^{-4})(1-\zeta^{-6})(1-\zeta^{-8})(1-\zeta^{-2})(1-\zeta^{-4})} \qquad (\zeta = e^{2\pi i/5})$$

$$\lim_{\varepsilon \to 0} \left\{ \sum_{s \in Z} \left(\frac{1}{i(s - 2/5)}\right)^{1+\varepsilon} + \sum_{s \in Z} \left(\frac{1}{i(s - 8/5)}\right)^{1+\varepsilon} \right\} .$$

Let

$$\zeta(u,\ a) = \sum_{n \geqslant 0} \frac{1}{(n+a)^u}\ , \quad a > 0,$$

be the Hurwitz's function. Then the series in the limit can be rewritten as

$$(-2 \sin \pi(1+\varepsilon)/2) \cdot (\zeta(1+\varepsilon,\ 3/5) + \zeta(1+\varepsilon,\ 2/5)).$$

It is well known that the Hurwitz's zeta function has a residue 1 at $u = 1$. It follows

$$N = -\frac{1}{5} \cdot \frac{\zeta^{-6k}}{(1-\zeta^{-4})(1-\zeta^{-6})(1-\zeta^{-8})(1-\zeta^{-2})(1-\zeta^{-4})}\ .$$

Take the trace of N in $Q(\zeta)$ over Q, we get the contribution from conjugacy classes in 7-10, 7-11, 7-14, 1-15, 1-18, 1-19, 7-22, 7-23 is

$$-5^{-2}[2,\ 0,\ -2,\ 1,\ -1].$$

Adding together, we get our assertion.

Combining Proposition 2 to Proposition 8 and contribution from regular elliptic elements of order 7, 9, 20 and 30, we get the following Theorem 1.

THEOREM 1. *For even integer* $k \geqslant 10$, *the dimension fromula for the vector space of Siegel cusp forms of degree three with respect to the modular group is given by*

$$\dim_{\mathbf{C}} S(k;\ Sp(3,\ Z))$$

$$= \text{Sum of Main Terms} + C_1(k)(2k-4)^2 + C_2(k)(2k-4) + C_3(k),$$

with $C_j(k+12) = C_j(k)$ $(j = 1,\ 2,\ 3)$. *Here the main terms are as shown in TABLE V.*

REMARK. This theorem gives the approximate value of $\dim_{\mathbf{C}} S(k;\ Sp(3,\ Z))$ up to a polynomial of degree two in k when k is greater than or equal to 10.

4.5 Determination of C_1, C_2 and C_3

To find the value of $C_1(k)$, we have to calculate contributions from conjugacy classes of elements which have a two dimensional set of fixed points. The process of calculation is carried out in [11] or can be done with similar arguements as previous section, so we only give the final results and omit the detail here.

1. $[e^{i\theta}, 1, -1]$, $\theta = \frac{\pi}{3}, \frac{2\pi}{3}$; (four conjugacy classes)

$2^{-9}3^{-4} \times (1+6) \times (2k-4)(2k-6) \times [1,-2,1]$, (leading terms).

2. $[\pm i,1,-1]$; (two conjugacy classes)

$(-1)^{k/2}2^{-12}3^{-2} \times (1+6) \times (2k-4)^2$.

3. $\begin{bmatrix} 1 & s \\ 0 & 1 \end{bmatrix} \times [1,-1]$; (three conjugacy classes).

$-2^{-11}3^{-3} \times (1+3+12) \times (2k-3)(2k-5)$.

4. $[\rho,\rho,\bar{\rho}]$, $(\rho = e^{\pi i/3})$.

$2^{-4}3^{-6} \times (2k-3)(2k-5) \times [1,-2,1]$, (leading terms).

5. $[i,i,-i]$; (two conjugacy classes)

$(-1)^{k/2}2^{-12}3^{-2} \times (2+4) \times (2k-3)(2k-5)$, (leading terms).

6. $[1,\rho,\bar{\rho}]$, $(\rho = e^{\pi i/3})$.

$2^{-5}3^{-4} \times (2k-3)(2k-5)$.

7. $[1,\rho^2,\bar{\rho}^{-2}]$, $(\rho = e^{\pi i/3})$, (two conjugacy classes),

$2^{-5}3^{-5} \times (1+12) \times (2k-3)(2k-5)$.

8. $[1,i,-i]$, (for conjugacy classes).

$2^{-11}3^{-2} \times (2+3+9+9) \times (2k-3)(2k-5)$.

Adding together, we obtain

$$C_1(k) = \frac{1}{2^7 3^2}[4,2,4,3,3,3]$$

$$+ \frac{1}{2^{12}3^6}[451,\ 1249,\ 451,\ 937,\ 763,\ 937].$$

REMARK. Let $P(k)$ be the sum of main terms. By soloving the system

$$\begin{cases} C_1(k)(2k-4)^2 + C_2(k)(2k-4) + C_3(k) + P(k) = m, \\ C_1(k)(2k+20)^2 + C_2(k)(2k+20) + C_3(k) + P(k+12) = n, \\ C_1(k)(2k+44)^2 + C_2(k)(2k+44) + C_3(k) + P(k+24) = p, \end{cases}$$

we get

$$C_1(k) = \frac{m-2n+p}{2^7 3^2} + \frac{1}{2^7 3^2}[2P(k+12) - P(k) - P(k+24)].$$

It follows

$$C_1(k) = \frac{1}{2^{12}3^6}[451,\ 1249,\ 451,\ 937,\ 763,\ 937]$$

$$+ \frac{1}{2^7 3^2}[m_1,\ m_2,\ m_3,\ m_4,\ m_5,\ m_6]$$

with integers m_1, m_2, m_3, m_4, m_5, m_6. This previous observation is consistant with our result in Theorem 2.

To determine the exact values of $C_2(k)$ and $C_3(k)$, we shall extract $C_1(k)(2k-4)^2$ and $C_\gamma(k)(2k-4)$ as shown in TABLE VII from the dimension formula so that the remaining terms can be write as

$$\tilde{C}_2(k)(2k-4) + \tilde{C}_1(k)$$

with $\tilde{C}_2(k) = C_2(k+6)$.

TABLE VII Some More Contributions

No.	Contribution	Conjugacy classes
1	$-(-1)^{k/2}2^{-11}3^{-1} \times 16 \times (2k-4)$	Parabolic elements with torsion part [i, 1, -1]
2	$-(-1)^{k/2}2^{-8}3^{-1} \times 4 \times (2k-4)$	Parabolic elements with torsion part [i, i, -i]
2	$2^{-6}3^{-2}$ $(2k-4) \times [1,0,-1,-1,0,1]$	Torsion elements with characteristic polynomial $(X-1)^2(X^2+1)(X^2\pm X+1)$.
	Sum $= 2^{-8}3^{-6} \times (2k-4) \times [-648, 972, -1296, 648, -972, 1296]$	

Now with data shown in TABLE I and TABLE III and $C_1(k)$ in Theorem 2, we can determine $\tilde{C}_2(k)$ and $C_3(k)$ up to a big constant which we can determine easily even by an obervation

PROPOSITION 9. Let

$$\tilde{C}_2(k) = C_2(k) + 2^{-8}3^{-6} \times [-648,972,-1296,648,-972,1296].$$

Then a general solution of $\tilde{C}_2(k)$ and $C_3(k)$ is given as follows:

$$\tilde{C}_2(k) = \frac{1}{24}[m_1, m_2, m_3, m_1, m_2, m_3]$$

$$+ 2^{-8}3^{-6}[-2362,-189,-3200,-2362,-189,-3200].$$

$$C_3(k) = [n_1, n_2, n_3, n_4, n_5, n_6]$$
$$+ 6^{-1}[-5m_1, -6(m_2+1), -7m_3, -2m_1, -3m_2, -4m_3]$$
$$+ 2^{-4}3^{-6}[-16070, 0, +6826, -1328, -2916, -5648].$$

Here m_i $(i = 1,2,3)$ *and* n_i $(i = 1,2,3,4,5,6)$ *are integers*.

Proof. We shall prove the case $k \equiv 0 \pmod{12}$. The remaining cases follow with a similar arguement.

Let the value of $\dim_{\mathbf{C}} S(12; Sp(3, Z))$ and $\dim_{\mathbf{C}} S(24; Sp(3, Z))$ be u and v respectively. Then

$$\begin{cases} 20\,\tilde{C}_2(12) + C_3(12) = -\dfrac{304360}{2^8 3^6} + u \\[2ex] 44\,\tilde{C}_2(12) + C_3(12) = -\dfrac{361048}{2^8 3^6} + (v - 14) \end{cases}$$

by a direct calculation. The solution of above system is given by

$$\begin{cases} \tilde{C}_2(12) = \dfrac{-u+v-14}{24} - \dfrac{2362}{2^8 3^6} \\[2ex] C_3(12) = u + \dfrac{5(u-v+14)}{6} - \dfrac{16070}{2^4 3^6}. \end{cases}$$

PROPOSITION 10. *The values* $C_2(k)$ *are given as follow:*

$$C_2(k) = \frac{-1}{24} + 2^{-8}3^{-6}[-3010, 783, -4496, -1714, -1161, -1904]$$

$$C_3(k) = 2^{-4}3^{-6}[5314, 0, 8770, 2560, 2916, 2128]$$

Proof. It follow from a direct calculation as shown in CHAPTER III.

REMARK. The approximate value of $C_2(k)$ is $-12\ C_1(k)$ by an observation of Eie's results [11].

REMARK. In the real computation, we note that $C_3(k)$ $(k = 0,2,4,6,8,10)$ are rather small rational numbers. In particular, we note that $C_3(2)$ is an integer. This forces $C_3(2) = 0$ by an obvious observation.

With previous propositions and Theorems, we then have

MAIN THEOREM I. *For even integer* $k \geqslant 10$, *the dimension formula for the vector space of Siegel cusp forms of degree three with respect to* $Sp(3, \mathbf{Z})$ *is given by*

$$\dim_{\mathbf{C}} S(k;\ Sp(3, \mathbf{Z})) = \text{Sum of Main Terms} + C_1(k)(2k-4)^2 + C_2(k)(2k-4) + C_3(k).$$

Here the Main Terms and values of $C_1(k)$, $C_2(k)$ *and* $C_3(k)$ *are as shown in TABLE V.*

The following two Theorems are already proved in [11].

MAIN THEOREM II. *For even integer* $k \geqslant 10$, *the dimension formula for the vector space of Siegel cusp forms of degree three with respect to the principal congruence subgroup* $\Gamma_3(2)$ *of* $Sp(3, \mathbf{Z})$ *is given by*

$$\dim_C S(k;\ \Gamma_3(2))$$

$$= [\,Sp(3,\ Z):\ \Gamma_3(2)] \times$$

$$2^{-15}3^{-6}5^{-2}7^{-1}(2k-2)(2k-3)(2k-4)^2(2k-5)(2k-6)$$

$$+ 2^{-15}3^{-4}5^{-1}(2k-2)(2k-4)^2(2k-6)$$

$$- 2^{-14}3^{-3}5^{-1}(2k-3)(2k-4)(2k-5)$$

$$- 2^{-13}3^{-3}(2k-3)(2k-5) - 2^{-14}3^{-2}5^{-1}(2k-4)$$

$$+ 2^{-13}3^{-1}(2k-4) - 2^{-13}3^{-1} + 2^{-13}3^{-3}] .$$

<u>Here</u> $[\,Sp(3,\ Z):\ \Gamma_3(2)] = 2^9 3^4 \times 35$.

<u>MAIN THEOREM III</u>. <u>For even integer</u> $k \geqslant 10$, <u>the dimension formula for the vector space of Siegel cusp forms of degree three with respect to the principal congruence subgroup</u> $\Gamma_3(N)$ $(N \geqslant 3)$ <u>of</u> $Sp(3,\ Z)$ <u>is given by</u>

$$\dim_C S(k;\ \Gamma_3(N))$$

$$= [\,Sp(3,\ Z):\ \Gamma_3(n)] \times$$

$$[\,2^{-15}3^{-6}5^{-2}7^{-1}(2k-2)(2k-3)(2k-4)^2(2k-5)(2k-6)$$

$$- 2^{-9}3^{-2}5^{-1}(2k-4)N^{-5} + 2^{-7}3^{-3}N^{-6}] .$$

Here $[\,Sp(3,\ \mathbf{Z}):\ \Gamma_3(N)] = \frac{1}{2}N^{21} \prod_{\substack{p|N \\ p \text{ prime}}} (1-p^{-2})(1-p^{-4})(1-p^{-6})$.

<u>REMARK</u>. For $N \geqslant 3$, it is well known that $\Gamma_3(N)$ is torsion free Furthermore, the characteristic polynomial of any element in $\Gamma_3(N)$ is $(X-1)^6$ by the reduction theory of U.

Charistian [6]. Thus any element M in $\Gamma_3(N)$ is conjugate in $Sp(3, Z)$ to an element of the form $[S, E_3]$, $S = {}^tS \in M_3(Z)$.

With the help of high-speed computer, we are able to write down explicit values of $\dim_{\mathbf{C}} S(k; Sp(3, Z))$, $6 \leqslant k \leqslant 118$, as shown in TABLE VIII.

TABLE VIII. Explicit Values of $\dim_{\mathbf{C}} S(k; Sp(3, Z))$

	k=12L	k=12L+2	k=12L+4	k=12L+6	k=12L+8	k=12L+10
L = 0				0	0	0
L = 1	1	1	3	4	6	9
L = 2	14	17	27	34	46	61
L = 3	82	99	135	165	208	261
L = 4	325	389	490	584	708	852
L = 5	1023	1200	1445	1687	1984	2327
L = 6	2717	3133	3663	4199	4838	5557
L = 7	6360	7225	8267	9344	10585	11968
L = 8	13489	15116	17037	19023	21271	23742
L = 9	26429	29324	32615	36050	39881	44047

4.6. <u>The partial fractions of the generating function</u>.

Let m be a positive integer with $m \geqslant 2$ and $\alpha(k) = [a_0, a_1, \ldots, a_{n-1}; 2m]$ be the periodic function defined by

$$\alpha(k) = a_j \quad \text{if} \quad k \equiv 2j \pmod{2m}.$$

Then the generating function of $\alpha(k)$ is given by

$$\sum_{k=0}^{\infty} \alpha(k)T^k = (\sum_{j=0}^{m-1} a_j T^{2j})/(1 - T^{2m}).$$

Suppose that $P(k)$ is a polynomial in k of degree r, then by an elementary conderation, we see that the generating function

$$\sum_{k=0}^{\infty} \alpha(k)P(k)T^k$$

appeared to be the form

$$(\sum_{j=0}^{mr-1} b_j T^{2j})/(1 - T^{2m})^r .$$

Note that the contribution corresponding to a conjugacy classes or a family of conjugacy classes appeared to be the form $\alpha(k)P(k)$, so that the generating function for moduler form of degree three, $G_3(T)$ can be written as

$$G_3(T) = (\sum_{j=0}^{54} a_j T^{2j})/(1-T^4)(1-T^{12})^2(1-T^{14})(1-T^{18})(1-T^{20})(1-T^{30}).$$

Here the denominator $(1-T^4)(1-T^{12})^2(1-T^{14})(1-T^{18})(1-T^{20})(1-T^{30})$ is obtained from the least common multiple of denominators in the generating functions for contributions.

To determine the coefficients in the numerator of $G_3(T)$, we need 55 independent conditions. Now we shall compute the number of conditions which can be obtained from our main terms as shown in TABLE I.

PROPOSITION 11. <u>The generating function for the contribution from conjugacy classes of elements with characteristic</u>

polynomials

$$(X^4+X^3+X^2+X+1)(X\pm 1)^2 \quad \text{or} \quad (X^4\pm X^3+X^2\pm X+1)(X^2+1)$$

or

$$(X^4\pm X^3+X^2\pm X+1)(X^2\pm X+1)$$

is given by

$$A_0(T) = (T^4-T^{10}-T^{24}+T^{26}-T^{28}+T^{30}-T^{34}+T^{36}-T^{38}+T^{40}$$
$$+T^{46}-T^{48})/5(1-T^{20})(1-T^{30}).$$

Proof. The contributions are given by

$$b_1(k) = 2^{-3}3^{-1}5^{-2}(2k-4) \times [1,0,-1,3,-3;10]$$

$$b_2(k) = 2^{-3}3^{-1}5^{-2} \times [-66,0,54,-54,66;10]$$

$$b_3(k) = \frac{1}{20}[1,0,1,1,-1,-1,0,-1,-1,1;20]$$

$$b_4(k) = \frac{1}{15}[1,0,1,0,0,-1,0,0,0,0,0,0,-1,0,0;30]$$

by our previous calculation.

The generating functions for $b_1(k)$, $b_2(k)$, $b_3(k)$ and $b_4(k)$ are given by

$$B_1(T) = (-4-4T^4+24T^6-36T^8+24T^{10}-16T^{14}+36T^{16}-24T^{18})/600(1-T^{10})^2,$$

$$B_2(T) = (-66+54T^4-54T^6+66T^8)/600(1-T^{10}),$$

$$B_3(T) = (1+T^4+T^6-T^8-T^{10}-T^{14}-T^{16}+T^{18})/20(1-T^{20})$$

and

$$B_4(T) = (1+T^4-T^{10}-T^{24})/15(1-T^{30}),$$

respectively. Add together, we get $A_0(T)$.

PROPOSITION 12. The generating functions for

$$\alpha_1(k) = 2^{-15}3^{-6}5^{-2}7^{-1}(2k-2)(2k-3)(2k-4)^2(2k-5)(2k-6),$$

$$\alpha_2(k) = 2^{-15}3^{-4}5^{-1} \times 31(2k-3)(2k-4)^2(2k-6)$$

$$\alpha_3(k) = -2^{-13}3^{-3}5^{-1} \times 16(2k-3)(2k-4)(2k-5),$$

$$\alpha_4(k) = 2^{-10}3^{-5}5^{-1}(2k-3)(2k-4)(2k-5) \times [-2,0,2;6],$$

$$\alpha_5(k) = 2^{-10}3^{-5}5^{-1}(2k-4)(2k-5) \times [-10,20,-10;6],$$

$$\alpha_6(k) = 2^{-9}3^{-5}5^{-1}(2k-4) \times [8,10,-18;6],$$

$$\alpha_7(k) = -(-1)^{k/2}2^{-12}3^{-2}5^{-1}(2k-4)^2,$$

$$\alpha_8(k) = -2^{-9}3^{-2}5^{-1}(2k-4),$$

$$\alpha_9(k) = \frac{1}{7}[1,0,1,0,0,0,0;14],$$

$$\alpha_{10}(k) = \frac{1}{9}[1,0,1,0,-1,0,0,-1,0;18]$$

are given by $A_j(T)$ $(j = 1, \ldots, 10)$ as follow:

$$A_1(T) = (1-7T^2+22T^4+42T^6+469T^8+413T^{10}+84T^{12})/2^9 3^4 5 \cdot 7(1-T^2)^7,$$

$$A_2(T) = 31(1-5T^2+11T^4+5T^6+20T^8)/2^9 3^3 5(1-T^2)^5,$$

$$A_3(T) = (60-240T^2+300T^4-504T^4)/2^9 3^3 5(1-T^2)^4,$$

$$A_4(T) = (60+60T^4-744T^6+3840T^{10}-5604T^{12}+5964T^{16}-4080T^{18}$$
$$+504T^{22})/2^9 3^5 5(1-T^6)^4,$$

$$A_5(T) = (-20-12T^4+4T^6+264T^8-204T^{10}-272T^{12}+312T^{14}-72T^{16})$$
$$/2^9 3^5 (1-T^6)^3,$$

$$A_6(T) = (-32-72T^4+128T^6+120T^8-144T^{10})/2^9 3^5 5(1-T^6)^2,$$

$$A_7(T) = (4-8T^2)/2^9 3^2 5(1-T^2)^2,$$

$$A_8(T) = (-1+2T^4+4T^6-9T^8+4T^{10})/2^8 3^2 5(1-T^4)^3,$$

$$A_9(T) = (1+T^4)/7(1-T^{14}),$$

$$A_{10}(T) = (1+T^4-T^8-T^{14})/9(1-T^{18}).$$

Proof. It follows from a direct calculation so we omit it here.

A least common multiple for the denominators of $A_j(T)$ $(j = 0, \ldots, 10)$ is $(1-T^4)(1-T^{12})^2(1-T^{14})(1-T^{18})(1-T^{20})(1-T^{30})$ which is precisely the denominator of $G_3(T)$ the generating function for modular forms of degree three. Now we may ask the question: How many conditions can be obtained from $A_j(T)$? This question is equivalent to determine the dernomiator of

$$G_3(T) - \sum_{j=0}^{10} A_j(T)$$

after a simplification.

To do this, we decompose $A_j(T)$ into partial functions with relative prime denominators. For example:

$$A_9(T) = \frac{1+T^4}{7(1-T^{14})} = \frac{2}{49(1-T^2)} + \frac{5+3T^2+8T^4+6T^6+4T^8+2T^{10}}{49(1+T^2+T^4+T^6+T^8+T^{10}+T^{12})} .$$

In the partial fraction of $A_9(T)$, the second term represents the only partial fraction in $G_3(T)$, which has denominator $(1+T^2+T^4+T^6+T^8+T^{10}+T^{12})$. Hence

$$G_3(T) - A_9(T)$$

can be simplified so that its denominator is $(1-T^2)(1-T^4)(1-T^{12})^2(1-T^{18})(1-T^{20})(1-T^{30})$. Consequently, 6 conditions can be obtianed from $A_9(T)$.

With a careful and elementary discussion, we get the following table for numbers of conditions can be obtained from $A_j(T)$ $(j = 0, 1, 2, \ldots, 10)$ and $1/(1-T^4)(1-T^6)(1-T^{10})(1-T^{12})$.

TABLE IX Conditions determined by $A_j(T)$ $(j = 0,1,2,\ldots,10)$

$A_j(T)$	Namber of conditions	Denominators of the partial fractions
1. $A_0(T)$	20	$(1+T^2+T^4+T^6+T^8)^2$,
		$1-T^2+T^4-T^6+T^8$,
$1/(1-T^4)(1-T^6)(1-T^{10})(1-T^{12})$		$(1+T^{10})+T^{20})/(1+T^2+T^4)$,
		$1+T^2+T^4+T^6+T^8$.
2. $A_1(T)$	2	$(1-T^2)^7$, $(1-T^2)^6$.
3. $A_1(T)$, $A_2(T)$	1	$(1-T^2)^5$.
4. $A_1(T)$, $A_2(T)$, $A_3(T)$	1	$(1-T^2)^4$.
5. $A_4(T)$	2	$(1+T^2+T^4)^4$.
6. $A_5(T)$	0	$(1+T^2+T^4)^3$.
7. $A_6(T)$	0	$(1+T^2+T^4)^2$.

(TABLE IX CONTINUED)

8.	$A_7(T)$	1	$(1+T^2)^4$
9.	$A_8(T)$	0	$(1-T^2)^3$, $(1+T^2)^3$
10.	$A_9(T)$	6	$(1-T^{14})/(1-T^2)$
11.	$A_{10}(T)$	6	$1+T^6+T^{12}$

From the above table, we see that 39 conditions can be obtained from $A_j(T)$ $(j = 0, 1, 2, \ldots, 10)$ with $1/(1-T^4)(1-T^6)(1-T^{10})(1-T^{12})$ and the rational function

$$G_3(T) - \sum_{j=0}^{10} A_j(T) - 1/(1-T^4)(1-T^6)(1-T^{10})(1-T^{12})$$

can be simplified so that its denominator is given by

$$(1-T^2)^3(1+T^2)^3(1+T^2+T^4)^3(1-T^2+T^4)^2$$

which is a polynomial of degree 32. Thus we need 16 conditions to determine $G(T)$ besides the conditions from A_j $(j = 0, 1, 2, \ldots, 10)$.

<u>REMARK 1</u>. The generating function obtained from the sum of generating functions for individual contribution is not correct for $0 \leqslant k \leqslant 8$. Thus a polynomial in T of degree no more than 8 has to add to the generating function. For example, we add $-1-T^4$ to the sum of partial generating functions in the case of degree three.

<u>REMARK 2</u>. To determine the explicit values of $C_1(k)$, $C_2(k)$ and $C_3(k)$, we need 16 conditions instead of 18 conditions. Indeed, only 4 conditions is needed in the determination of $C_1(k)$.

4.7. The generating function for modular forms of degree four.

In this section, we shall describe some partial results in our determination of generating function for modular forms of degree four. Though these results are still quite far away from an explicit generating function, they can provide certain information in our study of modular forms of degree four.

1. It is well known that the individual contribution appears to be the form $\alpha(k)P(k)$ with $\alpha(k)$ a periodic function and $P(k)$ a polynomial in k. The period of $\alpha(k)$ depends only on the eigenvalues of the torsion part of the corresponding conjugacy classes or the family of conjugacy classes and the degree of $P(k)$ can be determined by Hirzebruch's proportionclity principle. With the conjugacy classes represented by $(\pm E_6) \times \begin{bmatrix} 0 & 1 \\ -1 & 0 \end{bmatrix}$ as an example, we see that the period of the periodic function $\alpha(k)$ is 4 since $i^4 = 1$ and the degree of $P(k)$ is 6 since the dimension of the set of fixed points for these elements is 6.

Under this consideration, we see that the generating function $G_4(T)$ for modular forms of degree four appears to be the form

$$G_4(T) = \Big(\sum_{j=0}^{275} a_j T^j\Big) / (1-T^8)(1-T^{12})^2(1-T^{28})(1-T^{42})(1-T^{18}) \times (1-T^{36})(1-T^{10})(1-T^{20})(1-T^{30})(1-T^{60}) + P(T).$$

where $P(T)$ is a polynomial of degree $\leqslant 12$.

Consequently, 284 conditions is needed in the determination of $G_4(T)$.

2. There are only two contributions with degrees greater than 6: the contribution from identity E_8 and the contribution from conjugacy classes of elements of order 2. Let $\alpha_1(k)$ and $_2(k)$ be the contributions. Then

$$\alpha_1(k) = 2^{-24}3^{-8}5^{-4}7^{-2}(2k-2)(2k-3)(2k-4)^2(2k-5)^2(2k-6)^2(2k-7)(2k-8),$$

$$\alpha_2(k) = 2^{-22}3^{-7}5^{-2}7^{-1}\cdot 127(2k-2)(2k-4)^2(2k-5)(2k-6)^2(2k-7)(2k-8)(-1)^k.$$

The generating functions for $\alpha_1(k)$ and $\alpha_2(k)$ are given by

$$A_1(T) = (1-11T+55T^2-165T^3+330T^4-461T^5+493T^6-143T^7+495T^8 + 132T^9+42T^{10})/2^{14}3^5 5^2 7(1-T)^{11}$$

and

$$A_2(T) = [127(-1+8T^2-28T^4-T^5+70T^6-82T^7+203T^8-582T^9+896T^{10} -840T^{11}+554T^{12}-273T^{13}+90T^{14}-14T^{15})] /2^{12}3^5\cdot 5\cdot 7(1-T^2)^8.$$

3. Consider those elements with characteristic polynomials $(X\pm1)^2(X^6\pm X^3+1)$ or $(X^2+1)(X^6\pm X^3+1)$ or $(X^2\pm X^2+1)(X^6\pm X^3+1)$. These elements contain torsion elements of order 9, 18, 27 or 36 and parabolic elements with characteristic polynomials $(X\pm1)^2(X^6\pm X^3+1)$. Conjugacy classes for these elements can be classified by the reduction theory given in [6] and the corresponding contributions are given as follow:

(A) $2^{-3}3^{-4}(2k-5)\cdot[-1,-2,0,0,2,1,1,0,-1,2,1,0,0,-1,-2,1,0,-1;18] + 2^{-3}3^{-4}\cdot[-11,-6,0,0,-6,-11,-1,0,5,10,3,0,0,3,10,5,0,-1;18]$

(the contribution from elements of order 9 and 18).

(B) $\frac{1}{36}[1,0,0,0,0,1,-1,0,1,-2,1,0,0,-1,2,-1,0,1,-1,0,0,0,0,-1,1,0,-1,2,-1,0,0,1,-2,1,0,-1;36]$,

(the contribution from elements of order 36),

(C) $\frac{1}{27}[1,0,0,0,0,1,-1,0,1,0,0,0,0,0,0,1,0,-1;18]$,

(the contribution from elements of roder 27),

(D) $2^{-1}3^{-2}[1,2,0,0,-2,-1,-1,0,1,-2,-1,0,0,1,2,-1,0,1;18]$,

$+3^{-3}[1,2,0,0,-2,-1,-1,0,1,-2,-1,0,0,1,2,-1,0,1;18]$

(the remaining contributions).

Here $B(k) = [b_0, b_1, \ldots, b_{m-1}; m]$ denotes the function defined by

$$B(k) = b_j \quad \text{if} \quad k \equiv j \pmod m.$$

An elementary calculation shows that the generating function for these contributions is given by

$$A_3(T) = (1+T-T^4-T^6+T^8-T^9+T^{14}-T^{18}-T^{19}+T^{22}+T^{24}-T^{26}+2T^{27}-2T^{32} +T^{33}-T^{35}-T^{36}-2T^{37}+2T^{40}+T^{41}+T^{42}-T^{44}+T^{45}+T^{46}+T^{49} -T^{50})/9(1-T^{18})(1-T^{36}).$$

4. Consider those elements with characteristic polynomials $(X\pm 1)^2(X^6\pm X^5+X^4\pm X^3+X^2X+1)$ or $(X^2+1)(X^6\pm X^5+X^4\pm X^3+X^2\pm X+1)$ or $(X^2\pm X+1)(X^6\pm X^5+X^4\pm X^3+X^2X+1)$, we get the folloiwng contributions:

(A) $2^{-4}3^{-1}7^{-2}(2k-5)\times[2,-1,0,0,1,-2,0;7]$

$+\ 2^{-4}3^{-1}7^{-2}\times[-4,-3,0,0,-3,-4,0;7]$,

(the contribution from elements of roder 7),

(B) $2^{-4}3^{-1}7^{-1}(2k-5)\cdot[0,-1,0,0,1,0,0,0,1,0,0,-1,0,0;14]$

$+\ 2^{-4}3^{-1}7^{-1}\cdot[-2,-3,0,0,-3,-2,0,2,3,0,0,3,2,0;14]$,

(the contribution from element of order 14),

(C) $\frac{1}{28}[1,0,0,0,0,1,0,-1,1,0,0,-1,1,0,-1,0,0,0,0,-1,0,1,$

$-1,0,0,1,-1,0;\ 28]$

(the contribution from elements of order 28)

(D) $\frac{1}{21}[1,0,0,0,0,1,0,-1,0,0,0,0,1,0,-1,-1,0,0,1,0,0,0,0,$

$0,0,0,0,0,0,1,0,0,-1,-1,0,1,0,0,0,0,-1,0;42]$,

(the contribution from elements of order 21 or 42),

(E) $\frac{1}{14}[-1,1,0,0,-1,1,0,-1,0,0,0,0,1,0;14]$

$+\ \frac{1}{14}[0,1,0,0,1,0,0;7]$

(the remaining contributions).

With an elementary compatation, we see that the generating

function of these contributions is given by

$$A_4(T) = (T+T^5-T^7+T^8+T^{13}-T^{15}+T^{19}+T^{20}+T^{24}-T^{29}-T^{33}+2T^{35}$$
$$-T^{36}+2T^{39}+2T^{40}-T^{41}-2T^{43}-2T^{47}+T^{49}-T^{50}+2T^{51}+T^{55}$$
$$-T^{58}-2T^{61}-T^{62}-T^{63}+T^{64}+2T^{66}-T^{69})/7(1-T^{28})(1-T^{42}).$$

5. The generating function for the contribution from elements with characteristic polynomials $(X^2+1)(X^2\pm X+1)(X^4\pm X^3+X^2\pm X+1)$, i.e. elements of order 30 or 60, is given by

$$A_5(T) = (1+T^5+T^6-T^{15}+T^{19})/60(1-T^{10}+T^{20}).$$

6. The generating function for the contribution from elements with characteristic polynomial X^8-X^4+1, i.e. elements of order 16, is given by

$$A_6(T) = (1+T^5)/16(1-T^8).$$

7. The total contribution from conjugacy classes of elements of order 3,4 or 6 is given by

$$2^{-19}3^{-9}5^{-2}7^{-1}(2k-3)(2k-4)(2k-5)^2(2k-6)(2k-7) \times$$
$$[-91,-43,107,107,-43,-91,-37,11,53,53,11,-37;12]$$
$$+2^{-19}3^{-8}5^{-2}7^{-1}(2k-4)(2k-5)(2k-6)(2k-7)(4k-11) \times$$
$$[37,171,107,-107,-171,-37,91,-117,53,-53,-117,-91;12]$$
$$+2^{-18}3^{-9}5^{-1}7^{-1}(2k-4)(2k-5)(2k-6)(2k-7)\times$$
$$[416,305,-208,-529,-208,305,416,143,-208,-367,-208;12]$$

$$+2^{-18}3^{-9}5^{-1}7^{-1}(2k-4)(2k-5)(2k-6) \times$$

$$[497,-335,-945,-529,529,-945,335,-497,-783,-367,367,783;12].$$

Let $A_7(T)$ be the generating function for this contribution.

By the same argument as in previous section, we get the numbers of conditions which can be obtained from $A_j(T)$ $(j = 1, 2, \ldots, 7)$ as shown in the following Table.

TABLE X Conditions form A_j $(j = 1,2,\ldots,7)$

	$A_j(T)$	number of conditions	Denominators of the partial fraction
1.	$A_1(T)$	4	$(1-T)^{11}, (1-T)^{10}, (1-T)^9, (1-T)^8$.
2.	$A_2(T)$	4	$(1+T)^{11}, (1+T)^{10}, (1+T)^9, (1+T)^8$.
3.	$A_3(T)$	36	$(1+T^6+T^{12})^2, (1+T^6+T^{12}), (1-T^6+T^{12})$
4.	$A_4(T)$	60	$(1-T^{14})^2/(1-T^2)^2$, $(1-T^{14})/(1-T^2)$ $(1+T^{14})/(1+T^2), (1+T^{14}+T^{28})/(1+T^2+T^4)$
5.	$A_5(T)$	20	$1-T^{10}+T^{20}$
6.	$A_6(T)$	4	$1+T^4$
7.	$A_7(T)$	18	$(1+T^2+T^4)^7, (1+T^2+T^4)^6, (1+T^2+T^4)^5$ $(1+T^2)^7, (1+T^2)^6, (1+T^2)^5$

From the above table, we see that 146 conditions can be obtained from $A_j(T)$ $(j = 1, 2, \ldots, 7)$ and we can simply

$$G_4(T) - \sum_{j=1}^{7} A_j(T) - G_3(T)$$

so that its denominator is given by

$$(1-T^{12})^2(1-T^{20})^2(1-T^{30})^2(1-T^2)(1-T^2+T^4).$$

Here we shall like to write down a final remark for the dimension formula for modular forms of degree four with respect to Sp(4, Z). It is quite impossible to obtain the explicit formula by direct computations. However, with the help of high-speed computer and our knowledgement of modular forms, it is possible to obtain the generating function for modular forms of degree four.

REFERENCES

1. T. Arakawa, The dimension of the space of cusp forms on the Siegel upper half plane of degree two related to a quaternion unitary group, J. Math. Soc. Japan 33(1981), 125-145.

2. Teruaki Asai, The class numbers of positive definite quadratic forms, Japan J. Math. vol. 3, No.2, (1977), 239-296.

3. Walter L. Baily Jr., An exceptional arithmetic group and its Eisenstein series. Annals of Math. 91(1970), 512-549.

4. Z. I. Borevich and I. R. Shafarevich, Number Theory, Academic Press (1966).

5. Ulrich Christian, Zur Theorie der Symplektischen Gruppen, Acta Arithmetica XXIV (1973), 61-85.

6. U. Christian, A reduction theory for symplectic matrices, Math. Zeitschr. v.101, (1967), 213244.

7. Ulrich Christian, Unterschungeiner Poinaréschen Reine I, II. J. reine angew. Math. 233(1968), 37-88; 237(1969), 12-25.

8. Ulrich Christian, Berechung des Ranges der Schar der Schar der Spitzenformen zur Modulgruppe Zweiten Grades und Stufe $q > 2$. J. reine angew. Math. 277(1975), 130-154.

9. Ulrich Christian, Zur Berechung der Ranges der Schar der Spitzenformen zur Modulgruppe zweitzen grades und Stufe $q > 2$. J. reine angew. Math. 296(1977), 108-118.

10. Minking Eie, Contributions from conjugacy classes of regular elliptic elements in Sp(n, Z) to the dimension formula, to appear in Transaction of AMS (1984).

11. Minking Eie, Siegel cusp forms of degree two and three, to appear in Memoirs of AMS (1984).

12. Minking Eie and Chung-Yuan Lin, Fixed points and conjugacy classes of regular elliptic elements in Sp(3, Z). Manuscript (1984).

13. Minking Eie and Chung-Yuan Lin, Conjugacy classes of the modular group Sp(3, Z), Manuscript (1984).

14. R. Godement, Généralités sur les formes modulaires I, II. Seminaire Henri Cartan. 10e annes, 1957/1958.

15. Gottschling E., Über der Fixpunke der Siegelschen Modulgruppe. Math. Ann. 143(1961), 111-149.

16. Gottschling E., Über die Fixpunktergruppen der Siegelschen Modulgruppe. Math. Ann. 143(1961), 399-430.

17. Helgason S., Differential Geometry and Symmetric Spaces. Academic Press, (1962).

18. Ki-ichiro Hashimoto, The dimension of the space of cusp forms on Siegel upper half plane of degree two (I). Journal of fac. of science, University of Tokyo, vol. 30, No.2, (1983), 403-488.

19. L. K. Hua, On the theory of functions of several complex variables I, II, III. English translation. American Math. Soci. 32(1962), 163-263.

20. Jun-Ichi Igusa, On Siegel modular forms of genus two. American Journal of Math. 84(1962), 175-200.

21. R. P. Langlands, Dimension of spaces of automorphic forms, Proc. Symp. Pure Math. Vol. 9, Amer. Math. Soc. (1966), 253-257; Amer. J. Math. 85(1963), 99-125.

22. Hans Maass, Siegel's modular forms and Dirichlet series. Verlag-Spring, Lecture Notes in Math. 216.

23. George W. Machkey. Unitary group representation in physics, probability and number theory. The benjamin and Cummings Publishing Co., 1978.

24. H. Midorikawa, On the number of regular elliptic conjugacy classes in the Siegel modular group of degree 2n. Tokyo J. Math. Vol. 6, No.1,(1983),25-28.

25. Y. Morita, An explicit formula for the dimension of Siegel modular forms of degree two. J. Fac. Sci. The University of Tokyo, 21(1974), 167-248.

26. Y. Namikata, Toroidal compactification of Siegel spaces. Verlag-Springer, Lecture Notes in Math. 812.

27. Y. Namikata, A new compactification of the Siegel space and degeneration of abelian varieties. I. Math. Annalen 221 (1976), 97-141.

28. A. Selberg, Harmonic analysis and discontinuous groups in weekly symmetric Riemannian spaces with applications to Dirichlet series. J. Indian Math. Soci. 20(1956) 47-87.

29. Hideo Shimizu, On distinuous groups operating on the product of the upper half plane. Math. Annlen. 177(1963), 33-71.

30. Takuro Shintani, On Zeta-functions associated with vector spaces of quadratic forms. J. of Fac. Sci., The University of Tokyo. 22(1975), 25-65.

31. C. L. Siegel, Einfuhrung in die Theorie der Modulfunktionen n-ten Grades. Math. Annlen. 116(1939), 617-657.

32. B. Steinle, Fixpunktmannigfaltigkeiten symplektisher Matrizen, Acta Arithmetica, Vol. 20(1972), 63-106.

33. Ryuji Trushima, A formula for the dimension of spaces of Siegel cusp forms of degree three. American Journal of Math. 102(1980), 937-977.

34. Ryuji Trishima, On the spaces of Siegel cusp forms of degree two. American Journal of Math. 104(1982), 843-885.

35. S. Tsuyumine, On the Siegel modular form of degree three. Manuscipt, 1982.

36. T. Yamazaki, On Siegel modular forms of degree two. American Journal of Math. 98(1976), 39-52.

37. Lawrence C. Washington, Introduction to cyclotomic fields. Verlag-Springer, 1982.

CURRENT ADDRESS : Institute of Mathematics, Academia Sinica, Nankang, Taipei, Taiwan, Republic of China. 11529.

General instructions to authors for
PREPARING REPRODUCTION COPY FOR MEMOIRS

> **For more detailed instructions send for AMS booklet, "A Guide for Authors of Memoirs." Write to Editorial Offices, American Mathematical Society, P. O. Box 6248, Providence, R. I. 02940.**

MEMOIRS are printed by photo-offset from camera copy fully prepared by the author. This means that, except for a reduction in size of 20 to 30%, the finished book will look exactly like the copy submitted. Thus the author will want to use a good quality typewriter with a new, medium-inked black ribbon, and submit clean copy on the appropriate model paper.

Model Paper, provided at no cost by the AMS, is paper marked with blue lines that confine the copy to the appropriate size. Author should specify, when ordering, whether typewriter to be used has PICA-size (10 characters to the inch) or ELITE-size type (12 characters to the inch).

Line Spacing – For best appearance, and economy, a typewriter equipped with a half-space ratchet – 12 notches to the inch – should be used. (This may be purchased and attached at small cost.) Three notches make the desired spacing, which is equivalent to 1-1/2 ordinary single spaces. Where copy has a great many subscripts and superscripts, however, double spacing should be used.

Special Characters may be filled in carefully freehand, using dense black ink, or INSTANT ("rub-on") LETTERING may be used. AMS has a sheet of several hundred most-used symbols and letters which may be purchased for $5.

Diagrams may be drawn in black ink either directly on the model sheet, or on a separate sheet and pasted with rubber cement into spaces left for them in the text. Ballpoint pen is *not* acceptable.

Page Headings (Running Heads) should be centered, in CAPITAL LETTERS (preferably), at the top of the page – just above the blue line and touching it.

LEFT-hand, EVEN-numbered pages should be headed with the AUTHOR'S NAME;
RIGHT-hand, ODD-numbered pages should be headed with the TITLE of the paper (in shortened form if necessary).
Exceptions: PAGE 1 and any other page that carries a display title require NO RUNNING HEADS.

Page Numbers should be at the top of the page, on the same line with the running heads.

LEFT-hand, EVEN numbers – flush with left margin;
RIGHT-hand, ODD numbers – flush with right margin.
Exceptions: PAGE 1 and any other page that carries a display title should have page number, centered below the text, on blue line provided.

FRONT MATTER PAGES should be numbered with Roman numerals (lower case), positioned below text in same manner as described above.

MEMOIRS FORMAT

> **It is suggested that the material be arranged in pages as indicated below.**
> **Note: Starred items (*) are requirements of publication.**

Front Matter (first pages in book, preceding main body of text).

Page i – *Title, *Author's name.

Page iii – Table of contents.

Page iv – *Abstract (at least 1 sentence and at most 300 words).

**1980 Mathematics Subject Classification (1985 Revision).* This classification represents the primary and secondary subjects of the paper, and the scheme can be found in Annual Subject Indexes of MATHEMATICAL REVIEWS beginning in 1984.

Key words and phrases, if desired. (A list which covers the content of the paper adequately enough to be useful for an information retrieval system.)

Page v, etc. – Preface, introduction, or any other matter not belonging in body of text.

Page 1 – Chapter Title (dropped 1 inch from top line, and centered).

Beginning of Text.

Footnotes: *Received by the editor date.
Support information – grants, credits, etc.

Last Page (at bottom) – Author's affiliation.

[redacted]

ABCDEFGHIJ – 8987